Sylvia ... ton.
D1321146

Tunnicliffe's BIRDLIFE

Tunnicliffe's BIRDLIFE

CLIVE HOLLOWAY BOOKS
LONDON

For June Holloway

FRONTISPIECE: STORM PETRELS

These birds are rarely seen in Britain. Like all petrels they are oceanic birds and come to land only to breed and then usually at night. Moreover, although they breed in many places all round the northern shores of the Atlantic, and of the Pacific, in Britain they breed only in very remote and inaccessible islands of the far north-west. However, once in a while, usually after periods of protracted gales, these ocean dwellers appear close inshore and may be seen skipping along the line of breakers and, all too often, dead or dying on the shore. Tunnicliffe may well have seen this rare spectacle. He certainly had a specimen brought to him from which he made a measured drawing, no doubt used in the preparation of this oil painting.

First published in Great Britain in 1985 by
Clive Holloway Books,
48, Baldry Gardens,
London SW16

General text © Copyright 1985 by Noel Cusa

The paintings and sketches reproduced herein are
© Copyright of either the Estate of the late C.F. Tunnicliffe,
of the owner of the particular painting, or of a third party.

No part of this publication may be reproduced, stored in a retrieval system, or transmitted, in any form or by any means, electronic, electrical, chemical, mechanical, optical, photocopying, recording or otherwise, without the prior permission of the copyright owners.

British Library Cataloguing in Publication Data
Tunnicliffe C.F.
Tunnicliffe's Birdlife
1. Birds — Pictorial Works
I. Title   II Cusa, Noel
598'.022'2 QL 674

ISBN 0 907745 04 0

Designed by Nigel Partridge

Phototypeset by Bookworm Typesetting, Salford.
Printed and bound in Italy by New Interlitho, Milan.

AUTHOR'S ACKNOWLEDGEMENTS

The idea of this book was conceived some years ago by Robert Gillmor, who also at that time suggested that I should write a text for it. I should like to record my gratitude to him for starting off a project for which I share his enthusiasm.

Also I wish to thank Clive Holloway, who later, and independently, embarked on the same enterprise and pressed it to a conclusion.

I am grateful to Nicholas Hammond for bringing Clive Holloway and me together and to Clive for inviting me to join with him in carrying the exercise to fruition. It has been a pleasure to work with him and with his colleagues.

I am indebted to my wife, Mary, who has typed, re-typed, even in parts re-retyped, the whole of the manuscript without complaint and, almost, without error. She has done her best to prevent me from over-using favourite words and if the reader finds no monotony of repetition hers must be the credit. She has helped me vastly also in the sometimes painful task of selecting the best and discarding the almost-as-good and in contriving what we hope is a harmonious arrangement of the pictures.

Finally, if Charles Tunnicliffe is not too preoccupied in drawing the 'angels and archangels' of his new environment to notice what is going on here below I hope he approves of what we have done. It is his book, not mine. I think he would have been pleased with it. My gratitude to him for adding substantially to the manifold pleasures of my life is beyond words.

NOEL CUSA
RIVER HOUSE
LETHERINGSETT 1985

CONTENTS

FOREWORD BY SIR PETER SCOTT 9

INTRODUCTION 10

DUCKS AND RAILS 18

Mandarin Ducks 19
Mereside Waterfowl 21
Mallard and Magnolia 22
Moorhens 23
Shelduck Family 24
Summer Shelduck 25
Wigeon in Hard Weather 26
Teal 27
28 Pintail at Rest
29 Pintail Preening
30 Tufted Duck
31 Goldeneye
32 Eiders
33 Courting Pintails
34 Shoveler Ducks
35 Wigeon Grazing

GEESE 36

Gaggle at the Bar 37
Chinese Geese 39
Company of Whitefronts 40
Shelduck in Winter 41
Red-breasted Geese 42
43 Barnacle Geese
44 Alighting Whitefronts
45 Coming in to Land
46 Geese in the Morning
47 Canadas and Coots

SWANS 48

Mute Swans Preening 49
Rock and Roll 51
Mute Swan Flight 52
Bewick's Swans 53
54 Mute Swans Fighting
55 Whoopers Touching Down
56 Mute Swan Family
57 Black Swan

WADERS 58

Curlews and Turnstones 59
Snipe and Young 61
Green, Gold and Dun 62
Juvenile Ruffs 63
Curlews Alighting 64
Redshanks Alighting 65
66 Lapwing Family
67 Lapwing Flock
68 Black-tailed Godwits
69 Bar-tailed Godwits and Dunlin
70 Night Herons
71 Grey Herons

GULLS AND TERNS 72

Common Terns 73
Sunlight and Shadow 75
Gull Gallery 76
Ringed Plovers 77
78 Black-headed Gulls
79 Wader Reflections
80 Sandwich Terns
81 Black-headed Gulls

AUKS 82

Puffin Colony 83

BIRDS OF PREY 84

Buzzard 85
Short-eared Owl 87
Peregrine 88
Buzzard in the Rain 89
90 Snowy Owl
91 Gyrfalcon
92 Goshawk
93 Lanner

FOWL AND GAME 94

Fowl in a Damson Tree 95
Cock in the Wind 97
Sparring Cockerels 98
Ring-necked Pheasants 99
Blue-Eared Pheasants 100
Lady Amherst's Pheasants 101
Grey Partridges 102
French Partridges 103
104 Roosting Turkeys
105 Black Grouse
106 Peacock and Consorts
107 Peafowl Roosting
108 White Turkey
109 Leghorn Cockerel
110 Reeve's Pheasants
111 Elliot's Pheasants

PIGEONS 112

Fancy Pigeons 113

BEE-EATERS AND MAGPIES 114

Carmine Bee-eaters 115
Spring Magpies 116
117 In the Thorn Tree

BIRD PAINTING 118

Shelduck at Rest 129
Pheasant Display 135
138 Midsummer Godwits
140 Little Owl

FURTHER READING 144

INDEX 148

ACKNOWLEDGEMENTS 149

'It is with the creation of a very different kind of beauty that this book will try to deal, – that of line and form and colour on paper or canvas; a work of art in fact which, we hope, will have its own particular claim to be beautiful, not because it has slavishly imitated the form and colour of the bird, but because it has used the bird and controlled it to create a new beauty.'

C.F. Tunnicliffe
'Bird Portraiture'
The Studio, 1945.

FOREWORD

The verdict of posterity in time to come is likely, I believe, to rate Charles Tunnicliffe the greatest wildlife artist of the 20th century. The great Swedish painter, Bruno Liljefors, of the previous generation, might have a similar claim. At any rate I have been greatly moved by the work of both artists.

Tunnicliffe was a brilliant draughtsman and a first-class naturalist with a loving eye for the English countryside – especially of Cheshire where he was born and of Anglesey where he came to live. His technical handling of many graphic media was nothing short of masterly, but it was his genius for composition and rhythmical design which made all his works supremely satisfying to behold; this includes the exquisite measured drawings he made of virtually every dead creature that came into his hands.

As a dedicated naturalist his work was always faithful to nature. Not for him the deliberate distortions of some trendy modern art. He remained true to the animals he was portraying, believing that their reality could not be improved upon by exaggeration. And yet his style of painting is instantly recognisable and personally characteristic. The hallmark of his individuality emerges from the impeccable composition of each picture as well as the sheer skill of the drawing.

Tunnicliffe illustrated, wholly or in part, 87 books and from these – mostly water-colours, engravings, scraper-board and line drawings – his work is best known. His oil paintings and larger water-colours are less familiar to the public, which makes the appearance of this book all the more welcome and timely. Its object is to reproduce, many of them for the first time, these large works which represent the summit of Tunnicliffe's achievement – the end-product for which his measured drawings were a beautiful means. As the artist exhibited regularly at the Royal Academy – as an Associate and later as an Academician – many of these pictures are now in private hands, and therefore comparatively little known.

It is a happy circumstance that Noel Cusa should have compiled this book, for he was a close friend of Tunnicliffe, and is himself one of the leading wildlife artists of today. On occasions the influence of his mentor can be seen in Dr Cusa's own fine and very characteristic paintings.

As another painter who has been greatly influenced by the work of Charles Tunnicliffe, I am especially pleased to have been invited to contribute the foreword to this book, and therein to honour the achievements of a very great artist.

PETER SCOTT
THE WILDFOWL TRUST
SLIMBRIDGE, 1985

The Rivals from Llanddwyn

INTRODUCTION

The output of an artist can be broadly divided into two categories according to the purpose for which the picture is intended. On the one hand work may be designed to be viewed when held in the hand; on the other it may be meant to be framed and hung on a wall. In the first category there are illustrations for books, magazines or newspapers. Such work is necessarily carried out on a small scale and much of it is unavoidably in black and white. Except in the case where an artist is illustrating a book or article that he has himself written, there is normally another person involved in the initiation and design, an author maybe, or a publisher. The artist's originality and individual purpose is subject to various constraints, so that illustration, while it may, along with teaching, provide the bread and butter of an artist's life, does not call forth the spirit and inspiration, the individual and personal untrammelled initiative, from which the very greatest works of art may be expected to arise.

Illustration does, however, result in the artist's becoming widely known to the reading public. Each work is multiplied many times and falls into a multitude of hands. By contrast, a large painting made for wall decoration will normally be exhibited in a gallery, seen briefly by the few who frequent the gallery, sold, if the artist be fortunate, and then disappear, perhaps for ever, from public view. It thus commonly happens that the illustration work of an artist becomes much more widely known and appreciated than does his more individual, and perhaps greater work, that comes eventually to adorn the walls of private houses or public art galleries and museums.

Such circumstances are very pertinent to a consideration of the life work of Charles Frederick Tunnicliffe.

The son of a village cobbler turned farmer Tunnicliffe was, in his boyhood before the First World War, a typical farmer's lad. From a very early age he became skilled in the crafts and chores of life on a small mixed farm in the East Cheshire hills, not far from Macclesfield. But he also had a marked aptitude for drawing. He seems to have spent much of his childhood leisure in making likenesses of the domestic animals and birds of his farm home; some of them on the newly whitewashed walls of shippon and byre. This inclination and aptitude, encouraged at school and accompanied by an enlightened parental indulgence, made him into an artist. School led, not back to the farm or to agricultural college, but to Macclesfield Art School and thence to London and the Royal College of Art. He emerged from this thorough training as a fully qualified and competent artist, skilled in many techniques; drawing and the use of the scraper-board, etching, engraving on wood, painting in water-colours and in oils. A master of the many crafts that serve to make the complete artist.

His earliest professional work on leaving college, while still living in London, was a series of etchings which were well appreciated by collectors until etchings became unfashionable and Tunnicliffe abandoned this medium. He married, moved back to Macclesfield and shortly, at the suggestion of his wife, Winifred, began a long association with Henry Williamson as illustrator of several of his books. From this time on Charles rapidly built up a reputation as an illustrator of books of natural history and other country matters. He started with wood engravings for Williamson's '*Tarka the Otter*', published in 1932. This kind of work became for many years his main occupation. He illustrated much of the published writing of many authors, including, besides Williamson, Alison Uttley, Ronald Lockley, Ian Niall, E.L. Grant-Watson, Norman Ellison and H.E. Bates. Most of this work was in black and white, in earlier days executed by engraving on wood and later in pen and ink on scraper-board.

Thus, at least up to 1974, when he was 73, Tunnicliffe's public reputation rested largely on his very distinguished work as an illustrator. During his time in Macclesfield and his early years on Anglesey, immediately after the Second World War, he earned his living almost entirely in this way, and became very well known as possibly the best illustrator of country books in Britain. At the same time he produced a number of larger works and exhibited them, in consequence receiving much professional recognition. It was probably on the basis of his early etching that he was elected to the Royal Society of Painter-Etchers and Engravers. His admission to the Royal Academy as an Associate stemmed largely from his exhibits of wood engravings, but he had also begun to show paintings in water-colours and it was these which led to his becoming RA. They formed the basis of his annual submission to the Manchester Academy (until about 1954), to the Royal Academy's Summer Exhibition, and to several one-man shows at the Tryon Gallery in London. The etchings and wood-engravings were the pride of his youth. The exhibition work of his maturity was almost exclusively carried out in water-colours, and I think it is on this great series of paintings that he would, in the end, have wished to be judged as an artist.

While these paintings were very well appreciated by a discerning few, they were never the subject of fashionable critical acclaim – or condemnation. They were largely ignored by the art critics of the day. I cannot remember a single reference to Tunnicliffe's contribution in any critical review of the Manchester Academy's annual exhibition. Nor do I recollect any comment on his work in any press review of the Royal Academy's Summer Exhibition. It has to be borne in mind that Tunnicliffe's working life was passed during a period in which contemporary art criticism concentrated on the newest and most extravagant experiments. In most of these a good representation of natural subjects was held to be of no merit, or, if it was good of its kind, it excelled only in a region which did not constitute the highest and greatest in contemporary art.

Nevertheless, Tunnicliffe's water-colours, largely disregarded by popular criticism though they were, always found a ready market whenever and wherever exhibited. I do not remember a Royal Academy exhibition in which his annual contribution over many years, of six large water-colours, was not sold quite early in the show and in many cases at the private view. At the close of the exhibition his pictures almost invariably disappeared into private hands. They became the personal delight of a small group of enthusiastic patrons. True, a number were made rather more widely available by publication as greetings cards but it is clear that most of his paintings, for one reason or another, remained unknown save to the appreciative few who frequented galleries where his pictures were to be seen.

In Tunnicliffe's lifetime not only did he receive little or no critical recognition but very little was written about the man or his work. There is an article about him by an admiring friend, Sydney Rogerson, in *The Studio* magazine in 1945. His work is mentioned in an article in the same magazine by Raymond Sheppard, himself a fine animal illustrator, in 1952. Charles personally contributed 'C.F. Tunnicliffe, Yeoman Artist' to the magazine in 1942 while '*My Country Book*' (The Studio, 1942) and '*Bird Portraiture*' (The Studio, 1945) are to some extent autobiographical. But the man himself remained in the background, an obscure industrious figure, ceaselessly busy in a corner of Anglesey from which he rarely emerged. He had, as he put it himself, 'emigrated' to Wales as soon after the end of the war as he could be released from his job of teaching drawing to the pupils of Manchester Grammar School. There he lived, far away from the centres of population and publication, gossip and intrigue of the world of art, in another country and across the water. It was Winifred, his wife, who travelled to transact business with societies, publishers, galleries and agents. Charles stayed in Anglesey wholly occupied in doing what he liked doing, and did best, assiduously drawing, painting or engraving in the pursuit of a livelihood.

Then, in 1974, the Royal Academy wished to mount an exhibition of his work at Burlington House, to honour one of their more senior Academicians. He had no accumulation of suitable finished exhibition work in his studio but thanks to (as the then PRA Sir Thomas Monnington put it) 'the thoughtful persuasion' of Kyffin Williams, also an RA and a friend and neighbour in Anglesey, he agreed to spare from his studio part of his reference collection of sketchbooks and measured drawings of bird and animal corpses. The exhibition was a considerable success and was seen by a great many people. It came perhaps as something of a surprise to the art world: the quality of the work, the design of the sheets of measured drawings and the sustained dedication of the years devoted to their production. They were greatly admired and some came to regard them as his greatest work.

Towards the end of his life, full of years, somewhat taken aback by the honours he had received, with failing eyesight and increasingly unable to work, he agreed to the publication in book form of material from the 50 or so sketchbooks and of a selection of the more than 300 measured drawings that he had amassed during his crowded working years. Sadly it was not until after his death, in 1979, that these were published. '*A Sketchbook of Birds*' (Gollancz, 1979) was followed by '*Sketches of Bird Life*' (Gollancz, 1981) and '*Tunnicliffe's Birds*' (Gollancz, 1984). They have undoubtedly served to supplement the 1974 RA exhibition and to extend public appreciation of Charles Tunnicliffe, the natural history illustrator, to include a wide knowledge, appreciation and understanding of his very private work in executing these exquisite drawings.

After his death it was proposed to sell, by auction, his studio reference collection. To that end an exhibition was held at Christie's of the sketchbooks and of all the measured drawings. That exhibition, in 1981, was more complete than the earlier show at Burlington House and called even greater attention to this aspect of Tunnicliffe's work. Public emphasis was further accentuated by excited correspondence in the press. There was something of a clamour that this great treasure should not be dispersed. Happily, at the eleventh hour, the auction was averted and the complete collection was acquired by Anglesey County Council and, it is hoped, will eventually be on public display.

In these several ways it has come about that the two aspects of Tunnicliffe's working life that have been most widely displayed are, firstly, his voluminous and distinguished work as an illustrator and, secondly, his private labours in producing a hoard of life-sized measured drawings and a multitude of sketches from life. It has been easily overlooked that, when left to himself, not working to order, in collaboration with none but producing art for art's sake, he made several hundred pictures for exhibition that were shown in galleries and quickly dispersed into private collections. And yet the highly and justly acclaimed sketchbooks and measured drawings were but a means to an end. They were produced to provide reference material for his work as an illustrator and, more especially, for his exhibition paintings. They were, as he described them himself, but the 'tools of his trade'. I remember vividly his telling me, laughing, during the RA exhibition, 'I can't work; they have taken away the tools of my trade!' But he was, nevertheless, working, and he was indeed delighted that the Academy had mounted the exhibition in his honour. More profoundly so, I think, than by the more formal decorations, the gold medal of the Royal Society for the Protection of Birds and, later, the Order of the British Empire.

This book sets out to rectify this unbalanced public view of the artist's work, to present Charles Tunnicliffe at his peak as a mature artist, painting because he wanted to paint, making pictures of subjects which had appealed to his sense of the interesting, inspiring and decorative. In this book are fine reproductions of a selection of his best paintings. They must be appreciated, along with his work as an illustrator, his etchings and wood-engravings and his collection of sketchbooks and post mortem drawings, in order that the man may be measured in his full stature as an artist.

Reared as he was, a country boy on a farm, his earliest work took as subject matter the farmyard and surrounding landscape; particularly the domestic animals to be seen there. This interest in farm animals continued throughout his life. Much of his illustrative work depicted animals. A proportion of his exhibition etchings, wood-engravings and paintings, even in later years, were of animal subjects, both domestic and wild; horses (particularly), cattle, sheep, pigs, dogs, cats (especially Siamese cats), deer, foxes, badgers and otters. But, compelled by considerations of space, animal paintings have been excluded from this volume. The pictures chosen are all of birds, wild and tame. From the time of Tunnicliffe's return to Cheshire from London the study of birds became an increasing obsession. Stimulated no doubt by his illustrations for Henry Williamson's '*The Peregrine's Saga*', first published in 1934, his sketchbooks show a progressive dominance by birds. Animals, landscape and botanical studies began to take a secondary place. By the end of the war he was writing and illustrating '*Bird Portraiture*', a subject on which he by then justly considered himself expert. His exhibition paintings had become, as to the great majority, pictures of birds. Indeed his move to Anglesey in 1974 was undoubtedly in large measure stimulated by an urge to place himself in an environment rich and various in its birds.

None of the pictures reproduced here has been published before in book form and most of them have not been seen by the public save in exhibitions of limited duration. Many have not been seen at all as they were painted to fulfil private commissions and passed directly from the artist's studio to the walls of the patron. With the exception of a very few works in oils the paintings were carried out in water-colours, and it is with Tunnicliffe's work as a painter of birds in water-colours that this book is primarily concerned.

Birds have an intrinsic beauty. It explains to some extent why watching birds has become such a popular and pleasant pastime. Much painting of birds concentrates on this beauty independently of, and in disregard of, the creature's surroundings. Emphasis tends to be placed on feather detail. As Peter Scott says in his '*Observations of Wildlife*', published in 1980, 'Pictures of birds or mammals which show every feather or hair seem to be most admired'; it sometimes seems that the bird-watcher who buys a picture wants a record of what the bird might look like if his binoculars had the powers of a microscope. Tunnicliffe was sensitive to this intrinsic beauty of birds and his measured drawings of bird corpses made just such a record in careful detail. But, as he himself says in '*Bird Portraiture*', 'It is with the creation of a very different kind of beauty that this book will try to deal, that of line and form and colour on paper or canvas; a work of art in fact which, we hope, will have its own particular claim to be beautiful, not because it has slavishly imitated the form and colour of the bird but because it has used the bird and controlled it to create a new beauty'. So, in a picture of birds by Tunnicliffe, we must consider not only the birds but the picture as a whole; as something with 'its own particular claim to be beautiful'.

In the first place, the bird is treated as part of a landscape, as part of its surroundings. The setting that, in some artist's work, so often amounts to no more than the necessary twig so that the bird may perch. For the setting *is* a landscape, albeit in many instances not the expanse to the horizon of the landscape painter, but perhaps just a few square yards of ground – a landscape nevertheless. For Tunnicliffe the environment was as important as the birds and he devoted as much care and knowledge to accurate and convincing representation of the setting as of the birds themselves. In these paintings we shall see studies of rocks with their incrustations of moss, lichen or seaweed; of sand with its ripples and pools; of plants, plants in flower, plants sere and yellow in winter, plants freshly green and budding in spring; grasses, twigs, branches, fungi and rotting wood, all exactly observed and beautifully drawn and painted. But especially we shall see water; water in all its forms and moods; sea water calmly rippling or violently churned into the froth and splash and spume of waves; water in still pools, exactly reflecting; water falling from the heavens as rain or snow; water frozen to ice in sheets or in the icy droplets of hoar frost – this last particularly, for Charles was always fascinated by the decoratively varied forms of frost – water delicately softening outlines as mist, water lying massy and thick on ground or branch as snow or vaporous above as cloud of ever various formation. It has been said that he would have been famous as a painter of water if he had never seen a bird.

Light and its effects will be studied. Tunnicliffe was observant of shapes and sizes and representation in two dimensions of the solid geometry of his subject. He was also excited by the play of light and its effects on apparent shape, on colour, and on our appreciation of the time of day or of the weather. Sunshine on a clear day with its sharply cut shadows and blue sky-reflecting water; dull days, with little light or shade; bright summer mid-days with near vertical sun. Late afternoons in winter with their long cold shadows. Wild days of tormented sea and dim light from dark cloud; night with sombre skies and light diffused and feeble. And, particularly, light as influencing colour; the blueing light from a cloudless sky, the yellow light seen through a translucent wing; the coloured light bounced back onto the underside of a bird from sand or water or weed. He would often remark that there is 'no such thing as local colour' and was at pains to represent colour as he saw it in a particular setting and not as matching, as it might be, a decorator's swatch.

We shall often see movement in his pictures. Not just movement of bird or of water, but movement of the air. He was absorbed by the decorative possibilities of moving air, of zephyrs, breezes, gusts and gales. He strove to paint pictures in which such movement was convincingly conveyed. Birds at rest arrange themselves in elegant groups and submit to easeful study, birds in motion call for more perceptive observation. Many of Tunnicliffe's paintings were of resting groups of birds on still days but he also moved boldly into regions where most artists fear to tread: paintings of shore birds in a gusty wind lifting feathers awry and whipping pools to wave and spume; pictures of cockerels caught by a sudden blast, discomfited by inverted feathers and tottering to retain a dignified stance as dead leaves swirl by; studies of birds in flight, not just figures in flight posture, stuck there for ever, never moving, but birds in veritable motion; birds landing and taking off: in short, birds not just sitting there but doing things; cocks fighting, pheasants sparring, hawks mantling over prey, paper-fragile gulls lifting over a big wave, birds feeding, displaying, tending their young. All so much more difficult than the merely resting subject, being dependent on shrewd observation of the active creature. And on inspired and skilful sketches of various stages of movement.

Finally, in Charles Tunnicliffe's paintings we should note the consummate design and composition, which again is an area where many who draw and paint birds remain incompetent or blind. The bird centred on the sheet, supported by a necessary perch, doing nothing; a faithful and detailed zoological study with little or no merit as a picture. Tunnicliffe laid great emphasis on composition in his conversation about pictures and was much more pleased with his own work when he had devised a design that satisfied him, regardless of the merits of the ornithological representation. He was acutely sensitive to design in nature, to happy juxtaposition, to groups and settings that provided a ready-made picture. Such combinations were immediately noted in a sketchbook. With minimal adjustment, the sketch records the arrangement that ultimately makes the picture. More usually, however, there was a measure of invention in his pictures, in which juxtapositions were contrived, the decorative qualities, say, of vegetation recognised as an echo of a feather pattern. In almost every case a stage in the development of a painting entailed the making of a tiny sketch in colour, sometimes a series of sketches, in gouache usually. In these, with little or no regard for ornithology and none for detail, the scheme for the picture is set out as a pattern of colours and shapes. A few of these are reproduced here and it will be seen that the aesthetic qualities of the final picture, as a design, are already vividly apparent in them.

Tunnicliffe had studied the work of all the great artists, not only that of the great animal and bird artists. Though he cultivated ornithological accuracy – and was, as his mentor Reginald Wagstaffe taught him to be, more than 'a bit careful about birds' – he regarded such meticulous attention to detail as no more than a minor part of painting. He thought little of the work of the great illustrators, Lodge or Thorburn for example, though Audubon was held in the greatest respect. He greatly admired, and was influenced by, the Chinese and Japanese and by European painters of natural history subjects, especially the Swede, Bruno Liljefors, and the little known Englishman, Joseph Crawhall. He believed that a picture should be a pleasing aesthetic experience regardless of zoological content and one can see in his designs, in his patterns of colour, the influence of Gauguin and Matisse. Harmonies of line, shape, tone and colour were sought and studied with an inspired dedication. It is this quality of his work, superimposed on a faithful and accomplished skill in representation of bird and surroundings, that make Charles Tunnicliffe perhaps the greatest of bird artists.

The Rivals from Malltraeth Bay

DUCKS AND RAILS

There is no doubt that Tunnicliffe paid more attention to large birds than to small ones and, of the larger ones, he was perhaps most attracted to those of more strongly patterned plumage. Waterfowl of every kind were a constant source of inspiration and his early life in Macclesfield, when he pursued his birdwatching more often than not beside one or other of the Cheshire meres, provided him with abundant subjects among this group of birds. In his beautifully illustrated book '*Mereside Chronicle*' (Country Life, 1948) Tunnicliffe gives an account, in diary form, of his regular visits to these wonderful places and the wealth of waterfowl to be seen. His home in Malltraeth, in Anglesey was chosen in large measure because it overlooks the estuary of the river Cefni. The Cefni was at one time a meandering marshy river with high tides flooding far up the valley. In 1812 a sea wall was built, at Malltraeth, to retain the sea. The Cefni marshes were then reclaimed as farmland, the river being contained between embankments, with two flood water canals being provided. The sea wall is known as the cob (having been built with large shoreline boulders, or cobs), between it and the road there lies a lake which is known as Cob Lake or Cob Pool. This is partly fed by the river and partly, at high tides, by salt water from the sea. Thus it is a brackish water and for this reason, and to some extent, no doubt, due to the shelter provided by the Cob, it is a place greatly attractive to waterfowl notwithstanding the proximity of the road. Tunnicliffe got to know his neighbours, among them the shoregunners who produced many a waterfowl specimen to be studied, recorded and added to his collection of measured drawings. Regular watching and sketching from his studio window and, with a very short walk, at the Cob Pool, were supplemented by longer excursions to Llyn Coron, Cemlyn Pool and the estuary at Dulas. At all these localities waterfowl were the principal subjects studied.

The ensuing sixteen paintings are mainly pictures of ducks. Of waterfowl, ducks, or rather drakes, are probably the most colourful. Shelduck stand alone among British ducks in that male and female in that group are very nearly alike; in the majority of ducks the drake is a bird of gorgeous splendour whereas the female is usually a mottled and inconspicuous brown, designed by nature to be nearly invisible in her surroundings. Even drakes, however, reveal their distinguished finery only in a portion of the year, in some species for quite a short time. As soon as the breeding exercises are concluded the duck is left to her maternal duties and the drakes proceed to moult into what is known as an 'eclipse' plumage, more or less on the sober pattern of the female. This is retained for a shorter or longer period of late summer, autumn, and even early winter until, by a second moult, the breeding plumage is assumed in time for the ensuing nuptial activity. It is no accident, therefore, that of these paintings of ducks no fewer than seven depict winter scenes and some of the others show early spring. These are the seasons when drakes are at their best. In winter the British population of ducks is

Mandarin Ducks

Mandarins come from China and Japan and were introduced to Britain for ornamental waterfowl collections many years ago. In recent decades they have become established as a wild population in a number of places but particularly in and about Windsor Great Park. It has been suggested that they may be more numerous now in Britain than in their native lands, where they are reputed to be much reduced in numbers. The male is strangely adorned with crest and facial ruff and, most unusually, a 'sail' on each wing. Two drakes are shown displaying to the rather plain duck. In this display the special ornamentation of the drake is fully exhibited.

enormously increased by immigration from the Continent; the Cob Pool has visits from many more ducks and more varied species of duck, in winter and early spring than it does in summer.

It thus often occurs that in Tunnicliffe's paintings ducks are shown in a setting of patterned ice at the edge of a half-frozen mere or of frost on winter herbage; these settings were, in themselves, almost as fascinating to him as were the birds themselves. The pair of water-colours, one of Wigeon standing on ice, frozen, thawed and refrozen to concentric patterns, with emergent plants etched into natural design by hoar frost, and the other of Teal, similarly resting on patterned ice forms against a background of ochre reeds and dark willow, could well rank among his best pictures. They are certainly very characteristic of his work.

I have often heard Charles say that the Pintail was his favourite duck. He was always very conscious of the shape of birds and would refer disparagingly to the lumpy forms of Wigeon or to 'that great ugly black bill' of the Shoveler. He had nothing but appreciation for the elegant streamlined shape of the Pintail, which is apparent even in the female, though she lacks the needle tail of the drake and is, of course, very quietly coloured. The dark brown and white pattern of head and neck of the drake lends itself to design and three of the pictures Tunnicliffe painted of this elegant species are reproduced here.

Shelduck, of course, were comparatively rare on the inland Cheshire meres but were nearly always to be seen in the Cefni estuary. A creature of bold pattern – referred to by W.H. Hudson as 'the guinea-pig arrangement of black, white and red' – it was a favoured subject of Tunnicliffe's designs. Conscious of the somewhat bizarre clash between the crimson of the bill and the bright chestnut breast band, he was ever planning to turn it to advantage; but it is perhaps significant that in one of his best Shelduck pictures (page 129) the bill is visible in neither bird.

The Great Crested Grebe, a magnificent ornament of almost every Cheshire mere in spring and summer, appears in this collection only as a minor element in one picture where a pair is seen along with Tufted Ducks and Coot.

Mereside Waterfowl

This is a somewhat unusual picture for Tunnicliffe, suggestive of Cheshire rather than of his later Anglesey studies. The Coot, two pairs of Tufted Ducks and two Great Crested Grebes are small in relation to the picture size. The main area is occupied by a study of a sallow bush, its 'pussy' Easter catkins, the tracery of twigs and branches, and their reflections. The birds are seen through the tracery, the most striking of them, the grebes, all but obscured. The focus of interest is the black and white Tufted Ducks in the foreground. Very much the sort of view the bird-watcher gets, peering from the bank of a mere at a group of water-fowl.

Mallard and Magnolia

This painting of Mallard resting on a spit of grassy land by still water and seen through a screen of magnolia in bloom could well have been based on observation in St James's Park in London. It is perhaps more widely known than most of the paintings reproduced in this book, having been used some years ago to make a print that was issued as signed artists' proofs in limited edition under the title 'Mallard and Magolia'. Mallard are the most common of ducks but are by no means the least handsome. The drake with his yellow bill and dark, glittering head and breast would be much sought after were he but rare. The word 'mallard' originally implied the male but has been adopted as the official name for both drake and duck, both once known simply as 'wild duck'.

Moorhens

Moorhens (nothing to do with 'moors', but a corruption of 'merehen') are probably the most common of water birds in Britain. They are not critical in choice of habitat, provided it be wet, unlike the related Coot that demands a certain minimum of watery expanse. For a moorhen the merest trickle, tiny pool or roadside ditch seems to suffice. More solidly built and shorter of toe than the tropical jaçanas, the moorhen is nevertheless well able to skip along as they do over the pads of the more substantial water lilies. These chicks are tiny balls of animate sooty black fluff. The bare red skin about the face becomes more visible as they grow. A very successful species, the Moorhen has several broods a year and the feathered juveniles of the earlier broods often help to feed the later young.

Shelduck Family

Shelduck nest in burrows (indeed in some parts of Britain they are known as burrow geese) usually in soft sand. Dunes are thus a favourite nesting area and the first water seen by the newly-hatched ducklings is often, as shown here, a dune 'slack' or pool. In this interesting composition the rippled reflections of the adults have provided an effective camouflage for the boldly patterned and normally conspicuous young. What Hudson called the 'guinea-pig' arrangement of black, white and red makes the adults at all seasons very conspicuous but they, being reputed unpalatable, are rarely shot and so, happily, flourish.

Summer Shelduck

A Shelduck family rests in the sunshine by a sandy pool in the dune. Shelduck have some interesting habits. Before the young are fledged most adult Shelduck leave Britain for the vast sand flats of the Waddenzee off the German coasts, where they can moult, becoming temporarily flightless, in comparative safety. The young are left in their natal area, gathered into large groups or creches in the care of a few adults which moult locally. Nearly all the Shelduck of Western Europe behave in this way and fly to the same area of the German coast to moult.

Wigeon in Hard Weather

Wigeon are grazing ducks, and, often crepuscular, they leave the water at dusk to feed by cropping short grass. In daytime they usually rest and in hard weather can be seen standing about in ice and snow on the water surface or by the edge, where the patterns made by frost on rising and receding water among waterside plants attract the discerning eye of the artist. This picture is very much a picture of ice patterns and hoar frost in which the party of Wigeon provides a focus and the orange and chestnut heads of the drakes an accent of warmth in the general cold blue of birds and surroundings.

Teal

This is another picture of wildfowl in winter. This time the birds are Teal. Like the Wigeon opposite the ducks are standing on ice but there is no hoar frost on the vegetation and there is an unfrozen pool. The dark tree continues the arc of the group of birds. The sharp reflections in still water are contrasted with the hazy ones in ice. The bold white band on the side of the drakes is echoed in the rigid pale stems of the reeds, and the curved leaves sustain the pale cream lines patterning their vivid green and chestnut heads. It is a composition so subtle as to be, apparently, devoid of artifice.

Pintail at Rest

All the birds in this group of four Pintail drakes and a duck are asleep, with bill tucked into the feathers of the back. They are standing in still, shallow water and, save for the suggestion of rudimentary herbage at the bottom right, the ducks and their reflections constitute the design in the form of a diamond pattern in the picture rectangle. The elegance of the composition stems from the placing of the three white breasts and the two brown females, so that an otherwise too regular arrangement has that nicely judged asymmetry necessary to the design of a satisfying painting. Most artists would have thought it necessary that the birds show at least one head and neck complete, but how wrong they would have been.

Pintail Preening

This is a very fine painting of a pair of Pintail resting in calm shallow water. The drake is preening. The Pintail was one of Charles Tunnicliffe's favourite subjects. He was fascinated by the decorative potentialities of the boldy patterned head and neck, as indeed he was by those of Canada and Chinese geese. The duck is brown and hard to distinguish from the ducks of other species but the drake is very characteristic. The design is completed by the pale vegetation, carefully placed, and by the perfect reflections in the mirror-like water.

Tufted Duck

Tunnicliffe was fascinated by the constantly changing patterns of moving water and ingenious in suggesting them in paint. In this picture a flotilla of Tufted Ducks, six drakes and a duck, are swimming, probably on the Cob Pool, in a fresh breeze that has churned the shallow water into breaking wavelets among which the ducks bob along unperturbed. The female is the bird with the dusky flanks. The drakes have a more luxuriant drooping crest which harmonises happily with the curves of the rolling waves. Like the Goldeneye (opposite) both sexes of the Tufted Duck have golden eyes.

Goldeneye

Goldeneye are winter visitors to Britain, only a very few remaining to breed in recent years. But the wintering birds often linger well into spring and the nuptial display of the drakes is commonly seen. Here two drakes are in pursuit of a female and all three are about to alight on the rippled water. It is snowing. Goldeneye are quite often to be seen on the Cob Pool and they are equally at home in salt water as in freshwater. In flight the beating wings make a characteristic sound, the origin of the species name, *clangula*. They are diving ducks, but unlike most of their kind need no long run in order to rise from the water. They spring into the air as easily as Mallard.

Eiders

The Tunnicliffes frequently took their holidays in Scotland where the Eider is common. They are sea ducks and are rarely seen away from salt water. This picture shows a large flock rising and falling with the swell of a rough sea breaking on some coastal rocks. It is probably winter time, or very early spring, for the drakes are all in their nuptial plumage and the ducks have not yet dispersed to their breeding duties. In such gatherings there is usually much demonstration by the males and their musical 'hoo-hoo' note sounds eerily over the water. Tunnicliffe has made a grand picture by using the curves of the wave forms and of the flung spray where they break on the rocks.

Courting Pintails

Pintails are almost invariably to be seen in small numbers on the Cob Pool and although they do not breed in Anglesey they usually remain until well into spring and the communal display can be studied there. Here three drakes are paying attention to one duck in a typical posture with tail raised, head and neck drawn back and the bill depressed. The gently rippling water is summarily suggested and the lines of the wavelets are helpful in a nicely adjusted composition.

Shoveler Ducks

A picture of a resting flock of Shovelers, eight drakes and five ducks gathered on or near an islet of rough grass in a still pool. The grass is tawny and the drakes are in full breeding plumage so it is probably late winter or very early spring. The females merge with the ochre-brown grass and the diamond pattern is marked out by the drakes, whose black heads and white breasts are conspicuous in the fawn and blue setting. The lines of the huge spatulate bills and the margins of the white patches are repeated cleverly in the curving forms of the taller grasses.

Wigeon Grazing

In winter Wigeon are frequently found in very large flocks. In this somewhat unusual picture Tunnicliffe has departed from his customary picture organisation with a focus of interest and a careful balance of line, shape, tone and colour. Here he has produced a wallpaper-like all-over pattern of striking beauty and has suggested the large size of the flock not only by filling the picture rectangle but by showing some birds part in and part out of the frame. Veritably, as he himself entitled the picture, a 'Field of Wigeon'.

GEESE

Of the following ten pictures all but one are of geese. Geese were a favourite subject of Tunnicliffe's paintings. They are large birds and he liked large birds. They are shapely birds and he always had a eye for form and structure. During the time that he lived and worked in Cheshire he was able to study Canada Geese freely on the meres where they are plentiful. This bird is, in origin, a North American species. There, it behaves as the several species of migratory grey geese do in Europe. It nests in the far north and moves south to the United States in winter where it is a favourite target of sportsmen. It was introduced to Britain some 200 years ago as an ornamental bird, on lakes in the parks and gardens of the large landowners, the meres of the Cheshire and Shropshire plain are, or were, one of its main centres. In the British Isles it does not migrate, as its American ancestors did, but when breeding is over it gathers in large flocks in safe places to moult. At this time it is especially vulnerable, being flightless for some weeks. When the moult is over the flocks become restless and fidget from mere to mere but rarely move very far. The Canada Goose so prospered after the war that there were complaints from farmers of damage to crops, one remedy attempted was to catch the birds from the 'nuisance' flocks by driving them into pens while flightless. They were then transported to other, goose-less, parts of the country and in this way they have become much more widespread in Britain. When Tunnicliffe first moved to Anglesey there were no Canada Geese there but later some were transported to the island and released. They prospered, and Charles was able to continue to study and make pictures of these very handsome birds with their strikingly black and white patterned heads and necks and finely laced buff on brown bodies.

Lacking large river estuaries Anglesey is not one of the major resorts of wintering flocks of grey geese in Britain; but small flocks of Whitefronted Geese are usually there, together with a few Greylags. About the Cefni estuary, marshes and neighbouring areas there are often Whitefronts in winter-time and a gaggle flying past his studio window was no unusual winter spectacle for Tunnicliffe. These birds are the subject of several of the paintings in this section and I remember other magnificent pictures of Whitefronts in Royal Academy exhibitions. Large birds such as geese are more easily accommodated in landscape than small ones without giving rise to anomalies of scale and Tunnicliffe often departed from his more usual close-up style of picture in portraying them. Two of the pictures of the Cefni flock of Whitefronts reproduced here show them flying across a landscape recognisable as different views of the estuary near Tunnicliffe's home, 'Shorelands' at Malltraeth. I remember being shown 'Gaggle at the Bar' when it was newly finished, unframed in his studio, and its being explained to me exactly how the sky had been painted – with the paper upside down.

Chinese Geese are domestic geese of Chinese origin, derived from the Swan

Gaggle at the Bar

We are at the bar of the Cefni estuary and looking out to sea. It is dusk and the last of the sunlight is reflected in the sand pools. The geese are Whitefronts, though some of them are juvenile birds which lack the white 'front' and also, could we but see it, the irregular black stripes that adorn the bellies of the adults. Geese like to roost in the comparative safety of large sheets of water and 'flight' at dawn and dusk to their feeding grounds. If undisturbed they will feed by day but if harassed will do so at night. The morning and evening flights are the wildfowler's opportunity.

Goose of the Orient as European domestic geese are derived from the Greylag. They have long been introduced into Britain and can be seen in many parks (and even farmyards) in Cheshire. Their handsomely patterned heads and necks have decorative qualities they share with Canada Geese and Pintail. A design based on them was used on the dust cover of Tunnicliffe's book '*Bird Portraiture*'. In the picture we have chosen he has characteristically placed a group of them with necks raised against a pattern of hoar-frosted grass.

In some flocks of Chinese Geese the birds are wholly white with bill, legs and forehead knob orange. Tunnicliffe's sketches of these birds led to some notable pictures, a particularly memorable one showing two birds resting head to tail in a setting of mallow. This picture is reproduced in '*Portrait of a Country Artist*' by Ian Niall (Gollancz, 1980).

The paintings of Barnacle Geese and Red-breasted Geese will almost certainly have been derived from visits to collections of wildfowl. Barnacle Geese rarely come farther south in Britain than the Solway Firth and Red-breasted Geese are very rare stragglers indeed. Tunnicliffe's sketchbooks indicate at least one visit to the Wildfowl Trust collection at Slimbridge where he would have been able to study both species at ease. When he lived in Cheshire he regularly visited the pools that grace the grounds of Gawsworth Old Hall where there were a number of wildfowl, including Barnacle Geese. There are some beautiful duotone pictures of them in '*Mereside Chronicle*' the illustrated account of his Cheshire days.

Chinese Geese

This is a picture in which Tunnicliffe uses the decorative potentialities of Chinese Geese to advantage. Their boldly brown and white striped necks are stretched high and are held in elegantly clustered curves as the group of complaining geese moves away from some disturbance. These provide a natural focus for a picture of four birds in a setting of blue and white hoar-frosted grass, with which the darker blue-grey and white of their backs is harmonious. The bright orange of legs and feet is quietly repeated in the warm orange-brown of the breast.

Company of Whitefronts

Whitefronted Geese, winter visitors to Britain, come to escape from the savage winters of their summer home in northern Siberia. Here, at times, they find spells of almost equally arctic weather. Just such an occasion is the subject of this picture. Fourteen Whitefronts rest disconsolate in a land of hard frozen water and ground deep in snow. Tunnicliffe's picture effectively conveys the atmosphere of a still grey bitter day of hard frost. The pattern of snow and ice is skilfully organised as is the arrangement of the flock of geese. We may reflect how much better a picture it becomes by the presence of the broken foreground fence.

Shelduck in Winter

This is another sunless winter scene of ice and snow and cold water. It is an interesting picture (placed in this 'geese' section for aesthetic reasons only) in that it is scarcely typical of the artist. Tunnicliffe's designs are more usually crowded with figures, the picture rectangle broken into small areas of colour and tone to form a closely knit pattern. In this instance he has probably been influenced by the more sparse designs of oriental pointers, in which shrewdly placed small figures make a satisfying and elegant whole with a relatively large area of subtly varied background washes.

Red-breasted Geese

Red-breasted geese are bonny little geese that breed in Arctic Russia and winter principally around the Black and Caspian Seas. Apart from the odd bird being caught up in an east to west movement of other geese, usually Whitefronts, they are rare visitors to Britain. They are common birds of wildfowl collections and some reported occurrences in this country are doubtless due to escapes from these places. I doubt that Tunnicliffe ever saw a wild Red-breasted Goose and it is probable that this painting is an inventive composition, based on sketches made from captive specimens. The somewhat bizarre pattern of black, white and red in the individual birds is skilfully submerged in the arrangement of the group to make a very pleasing whole.

Barnacle Geese

Barnacle Geese are regular visitors to Britain but only rarely to Cheshire or to Wales. There are three breeding populations, in East Greenland, Spitzbergen and Novaya Zemblya. The Greenland birds winter in Ireland and the Hebrides, the Spitzbergen ones on the Solway and those from Russia in the Low Countries. The geese in Tunnicliffe's picture are evidently resting on the machair and white sands of the Hebrides and are, presumably, birds from Greenland. I do not know whether or not he visited wintering flocks of Barnacle Geese but his sketchbooks have numerous drawings made from birds in collections of wildfowl. A painting of Barnacle Geese called 'Solway Company' was, I think, one of the pictures he submitted for election as an Associate of the Royal Academy.

Alighting Whitefronts

Here the Whitefronted Geese are coming in to land on the sedge marsh down the estuary from Tunnicliffe's home. The dunes of Malltraeth Point are on the left and on the right is the rocky shore of Pen y Parc. The tide is out and the line of breakers is in the far distance. A flock of gulls rests on the wet sands of the estuary. The geese are swinging down to join others already standing beside the tidal pools among the sedge. These have not yet begun to feed and their vertically held necks are reflected in the pool and echo the down-tilted wings of the rapidly descending birds.

Coming in to Land

In this picture Whitefronted geese are coming in to land on Malltraeth beach. Llanddwyn Island and its white lighthouse tower are seen beyond and in the distance are the blue shapely forms of Yr Eifl, the Rivals, on the Welsh mainland across from Anglesey. In the foreground are the remains of a wrecked ship, long ago sunk and washed ashore and now all but buried in the sand. The picture gives a grand impression of such a scene on a clear, fine day in winter.

Geese in the Morning

Tunnicliffe made several paintings on this theme with a group of Canada Geese moving among hoar-frosted waterside vegetation. It combines two favourite subjects, Canada Geese with their boldly black and white patterned heads and necks; and the infinite variety of beautifully decorative forms to be perceived in the battered vegetation of winter, when picked out by an incrustation of rime after a very cold, humid night. In this instance the mists of early morning have dispersed to a brilliantly clear blue day in which sunshine has not yet melted away the fragile beauty of the frost. In other pictures on a similar theme the artist made use of early morning mist to give a dreamy recession to the goose flock.

Canadas and Coots

The decorative qualities of the black and white heads of Canada Geese and necks constantly attracted Tunnicliffe. In this picture the necks of the three right-hand geese provide the focus and in the foreground the black Coot serve to balance the composition; the fourth goose by its posture provides a needed link between the two dark areas. The golden vegetation harmonises beautifully with the brown of the backs of the geese, and the pale edges of the feathers of back and flank are continued in the pale stalks of the sere herbage.

SWANS

There are three kinds of white swan to be seen in Britain. The well-known Mute Swan is present in Anglesey throughout the year as well as being a common inhabitant of the Cheshire meres. The other two white swans, the Whoopers and Bewick's, are winter visitors. Bewick's Swan comes regularly to Anglesey and a flock of twenty or so of them were often to be seen on the Cob Pool. Whoopers are rather less common but do occasionally turn up on the Cob Pool or the Cefni estuary, so that Tunnicliffe had opportunity for intimate study of all three species. Whoopers and Bewick's Swans are very similar and until the middle of the last century were thought to be of the same species. It was Yarrell who demonstrated that the wintering wild swans in Britain were of two species and he called the newly distinguished one after Thomas Bewick, the wood-engraver who was author and illustrator of '*A History of British Birds*' published in 1832. Bewick's Swans are to be seen in their hundreds at the Wildfowl Trust's reserve at Slimbridge which Tunnicliffe had visited to supplement sketch references he had collected elsewhere.

The Black Swan is not a native and is known in Britain only as a captive bird. Tunnicliffe's painting is developed from a sketch he made (during a visit to the Wildfowl Trust at Slimbridge) of what the staff referred to as 'Winston Churchill's Swans', resting on a bank scattered with the fallen petals of a pink cherry tree. He also had several pencil sketches of individual birds in which he carefully studied the form of the curiously twisted wing feathers.

The Bewick's Swans that wintered on the Cob Pool at Malltraeth were of course the subject of continuous attention. The picture we reproduce, 'Heavy Weather', is an imaginative composition of these swans flying against a wild sea. The other swan paintings are of the common Mute Swan. He was attracted to the elegant shape of this bird, especially in the 'busking' display of the male or 'cob', and by the play of light on its white plumage. He has a chapter in '*Bird Portraiture*' in which he discusses the way in which a 'white' bird so rarely appears to be really

Mute Swans Preening

A female Mute Swan, the 'pen', and three cygnets have come ashore and, deep in dry sedge and other plants, are busily preening. The youngsters are well-grown and feathered but are still downy about head and neck. Tunnicliffe has evidently been struck by the decorative qualities of the natural arrangement of four curving necks. He has made little adjustment to the group as noted in his sketch, made at Cemlyn Little Pool (page 131), save to move the white adult to a more central position and to add the gorse bush as a dark contrast to the white bird. He called the picture 'Preening in the Sedges', and it was exhibited at the Royal Academy.

white and indeed varies in both colour and tone with its environment. It may be salutary to quote some of it:

Notice the yellow tinge in the feathers of neck and upper breast, and the cold bluish purity of the back, wings and tail. Note also the colour of the shadowed under-surfaces, and how it is influenced by the colour of the ground on which the bird is standing: if he is standing on green grass, then the under-parts reflect a greenish colour, whereas if he were on dry, golden sand, the reflected colour would be of a distinctly warm tint; or again if he were flying over water his breast, belly and under-wings would take on a colder tint, especially if the water were reflecting a blue or grey sky.

Or, when the swan is swimming in sunshine among dark reflections:

The bright light turns his back to a dazzling whiteness in which there is a tinge of gold or yellow; the shadows on his plumage are now blue to violet until they reach the under-sides on which you will see much reflected light from the water. Look especially at the colour of the shadows cast by the feathers of the raised plumes of the wings; you will see that the sun is able to penetrate through the vanes of the plumes and, in their depths, the shadows are of a different hue from those on the body of the swan, being greenish – even golden green.

Again, when the swan moves away from the dark reflections into the bright sky-reflecting water:

Soon he swims into a position in which both he and the sun are directly in front. In this view only his upper surfaces are lit by sunlight, the rest of him being in shadow and appearing dark violet against the bright water; in fact, but for the light on his back and the top of his head he appears as a dark silhouette in relation to the high tone of the water.

Or, in flight:

In front of a dark bank of trees he looks a very white bird but the moment he is above the trees and has the sky for a background he again becomes a dark shape with only his upper parts catching the light.

Or, in snow:

Now you can see how yellow his neck is and, to a lesser extent, the rest of his upper plumage. Note also the reflected snow light on his under-sides which makes them look almost the same tone as, or even lighter than, his top surfaces.

These observations serve to enlarge on his dictum that 'there is no such thing as local colour'. This is true of any solid body whether bird or not.

Rock and Roll

A family of Mute Swans, pen in the foreground, cob to the rear, three feathered youngsters between, is swimming against the current on choppy water with waves rolling towards the shore in a brisk breeze. The scene was probably observed on the Cob Pool. As they swim the birds are lifted and dropped in a rocking motion as the waves roll under them. Another of Tunnicliffe's distinctive titles, 'Rock and Roll', fairly describes the curious effect. The cygnets are in the grey-brown of the first feathering with patches only of the pure white of adult plumage; there is no sign of yellow or orange on the pinkish-grey and black bills.

Mute Swan Flight

It is a fine day of clear light, probably after recent rain. The embankment, known as the Cob, is in the foreground and beyond it lies the Cob Pool. Beyond the pool Bont Farm may be seen and the rolling sunlit farmlands of south-western Anglesey. In the distance the mainland hills rise to Yr Wyddfa, Snowdon in English. Overhead seven swans fly over the Cob and out towards the Cefni estuary. They are 'mute' in contrast with the bugle calls of the 'wild' swans of winter but in flight the beat of their great wings produces a noble sound. A fine painting of the scene near Tunnicliffe's home. Note the echo of the shapes of the darker shadows under the wings of the swans in the dark bushes on the Cob.

Bewick's Swans

Tunnicliffe's title for this picture was 'Heavy Weather'. It is a dark, wild day in winter. A party of seven swans flies low over a rough sea. These swans are winter visitors from Arctic Russia. Very similar to Whoopers, they are rather smaller and shorter of neck, but distinguished most readily by the smaller area of yellow on their otherwise black bills. Tunnicliffe has made a grand picture of an impressive scene, the squally sea as convincingly presented as the birds.

Mute Swans Fighting

Tunnicliffe's title for this picture was 'The Rivals' and we are to suppose that it represents pursuit of one male, or cob, by another rather than attention by the cob to the female, or pen. However it is hardly possible to tell certainly, with so obscure a view of the fleeing bird's bill. The black knob at the base of the bill is much larger in the cob than in the pen, especially in the breeding season. Mute Swans mate for life and although they are gregarious outside the breeding season the pairs occupy and defend a territory, albeit sometimes quite a small one, when nesting. Trespass by neighbouring swans leads to conflict, the intruder normally accepting defeat and retreating beyond the territorial boundary.

Whoopers Touching Down

This picture shows two Whooper Swans alighting on still water. The patch of yellow on the bill extending beyond the nostrils marks them as Whoopers rather than Bewick's. Like Bewick's Swan, the Whooper is a winter visitor to Britain from the north. The characteristic attitude of the alighting swan is well displayed, wings full spread, neck raised, legs out-thrust and webs spread to control the impact. Note the shrewd observation of varying colour and tone in the shadowed parts of the birds and the skilful representation of the just-disturbed water, not only the birds but the splash reflected in the mirror surface.

Mute Swan Family

The pair of Mute Swans (above) is idling on a reed bed. The now disused nest is in the corner of the picture and the four very newly hatched cygnets are ensconced on the pen's back. The cob behind (with the larger bill knob) is in the characteristic 'proud father' attitude, a warning threat to all intruders. This behaviour is known as 'busking'. The secondary feathers of the wings are raised in an arch over the back and the neck is curved back between them. Upon sighting an intruder the cob would move forward to the attack with a special stroke using both feet together (compared with normal swimming using alternate strokes) in an impressive surging motion, ploughing up a crest of water before him, hurling abuse with hisses and grunts despite his 'mute' nomenclature.

Black Swan

Black Swans (shown in the picture on the right) are natives of Australia. They have been introduced into New Zealand and have become common there where, curiously, the Mute Swan, which has also been introduced, does not seem to flourish. In Britain Black Swans have for long been kept for the decoration of ponds and lakes on private estates and in collections of wildfowl. Doubtless many are full-winged but the Black Swan is unknown as a wild bird here. The cygnets are newly hatched. The family party rests on a bank under cherry trees in bloom. The fallen petals of the cherry blossom extend the focal interest of the crimson bills of the adults, and the trunks of the cherry trees help the pattern of necks and legs.

WADERS

My wife and I once called on Tunnicliffe shortly after he moved into 'Shorelands', his bungalow in Malltraeth, and found him busy putting a layer of concrete all over his front garden! It was a hot day, he was stripped to the waist and his bulky torso was shiny with sweat and pink from the bright sunshine. Immediately over the garden wall was the estuary. The tide was low and sand and mud and rock were exposed among the pools. It was late summer and the waders were migrating. Just over the wall were Redshanks, Dunlins, Ringed Plovers and Bar-tailed Godwits. I remarked on these and I remember he said, 'Yes, and I would rather be drawing them than mixing concrete.' It was amusing to think of the artist putting his garden down to concrete so that he need no longer work on it. But I suspect the real reason was that salt sea spray over the wall at high tide in a south-westerly gale meant that little would grow there anyway. Besides, there was still a large garden at the back of the house that was well cultivated – largely by Winifred I think.

The constant proximity of waders just over the garden wall provided an endless variety of subjects. Only a short distance away the Cob Pool was a fine place to study the various species which were also to be found across the road on the field pool of Bont Farm lands. Black-tailed Godwits frequented the Cob Pool in spring and again in late summer, Bar-tailed Godwits came to the estuary in autumn; in April on Newborough Warren one might run across the handsome northern race of the Golden Plover. All of these received special attention. Rarities turned up on migration and were duly recorded and drawn in the sketchbook, phalaropes, stints, and strange waders from America. I remember once being shown a drawing of what I had little doubt was Baird's Sandpiper but Charles insisted it remain a mystery – so well had he digested his friend Wagstaffe's teaching to be 'a bit careful about birds'. But it was the common species of constant familiarity to which he usually resorted for his wader pictures, Oystercatchers, Redshanks, Ringed Plovers, Dunlin, Lapwings; and the resting parties of mixed waders that form such pleasing groups.

The pictures of flying waders alighting are interesting. They are relatively small birds and, unlike geese and swans, do not call for a vast landscape to accommodate them. Even in flight they will usually be represented in only a small patch of background. But in, for example, the painting of Redshanks Alighting (page 65), Tunnicliffe has chosen to present them as features of a landscape in the more accepted sense; in which the prospect before the observer extends to the horizon and beyond. Indeed in this picture we have, as in some of the pictures of geese and swans, truly figures in a landscape. If we can imagine the birds removed from the picture there remains a very fine landscape painting; of a quality that suggests that, had he so chosen, he could have become equally eminent as a landscape artist.

Curlews and Turnstones

Eight Curlews feed on a beach among the sand pools left by the tide. A rock, decorated with barnacles and seaweed, extends across the picture and three Turnstones are resting on it. In the distance the gentle waves of a calm sea rise and fall. It is probably late summer or autumn, when Curlews have left their marshy moorland nesting localities, gathered into flocks and come down to the shore to feed. Likewise the Turnstones, adults already in winter plumage, or maybe juveniles, have already come far south from their sub-arctic nesting areas. This is just such a group as may often be seen on Anglesey shores and in Tunnicliffe's picture the browns and greys are woven into a very subtle and satisfying pattern.

The upper quarter of the picture is occupied by a sunlit sky of cumulus clouds and the line of the headland is clear and darkly blue, as on a day of brilliant sunshine after showers such as are frequent in late summer in Anglesey. The sea is dead calm. No line of breakers is to be seen. The pools in the sands are quite still and as exactly reflecting as a mirror might be. The pattern of sand, water and sky is elegantly contrived but, if we continue to exclude in our mind's eye the alighting flock of birds, we may perhaps suppose the birdless landscape to be less satisfying for want of vertical contrast to the essentially horizontal pattern, relieved only by the bulbous forms of the clouds.

The birds are in two groups. A lower group of four have all but alighted, their long legs at full stretch forward as they are about to touch down, their wings beating downwards to control descent and the reflection of their legs, as it were, rising to meet them as they approach the water surface. Seen from above the bold dark brown and white pattern of the upper wings provides the focus of the picture, the white arcs of the wings echoed subtly in the curves of the cloud forms above. The upper group of five birds is more tightly bunched and indeed one bird is all but hidden from view behind two others. Not yet on the point of alighting, the legs of this group extend backwards, only just beginning to drop from the tight under-tail posture of level flight. Little is seen of the white arcs of the upper side of the upheld wings as they neatly break the possible monotony of the horizon line. Finally, the mandibles of at least four of the birds are apart, indicative of the noisy piping characteristic of such a troop coming to land after, perhaps, being disturbed from another feeding place.

By careful and constant observation of such flocks in flight Tunnicliffe has been able not only to make a static design of studied elegance but also to convey irresistibly an impression of rapid movement and bubbling sound, the more convincing by reason of the peaceful stillness of the surrounding landscape. Paintings of flying birds are often disappointing because the figures remain frozen in place, always the same place whenever the picture is seen. In this picture, such is the skilful translation to paint and paper of penetrating observation that it is difficult to believe that the birds in fact do remain in the same place.

We have included Herons conveniently among waders but, though they undeniably wade, no ornithological classification would group them thus. Two paintings of Herons are reproduced, one of Night Herons, the other of a pair of the common Grey Heron of Britain. Doubtless the Night Heron painting was derived from zoo studies but the common Heron is indeed common and was always to be seen about the estuary and the Cob Pool. Tunnicliffe liked these sometimes gawky birds. With their long legs and necks they fall naturally into the panel format and he found their decorative qualities very attractive.

Snipe and Young

This must be a scene that few have witnessed. Snipe usually nest in swampy ground, with dense vegetation such as this. The young are among the most attractive of wader chicks at the downy stage and, like all their relatives, crouch invisibly. It is unusual to see an adult with four young. Curiously, almost immediately after hatching, the young are shared, two and two, between the parents. The two halves of the family then live quite independently. In this picture the marsh plants, orchids, marsh pennywort, sphagnum and sedges are beautifully suggested.

Green, Gold and Dun

This is very much a bird-watchers' picture, developed from an interesting group of birds that Tunnicliffe saw and recorded in his sketchbook. Not, in his own estimation particularly distinguished as a composition, it does nevertheless effectively convey a particular scene and the low, soft light of a dull February day. The exact and careful rendering of the mirror-like reflections is a tribute to more than just shrewd observation. Tunnicliffe's title, 'Green, Gold and Dun', is interesting. It recalls that Lapwings are often called 'Green Plover' and that the name 'Dunlin' (meaning 'little grey one') is very apt to this common wader in winter plumage. In breeding plumage they are quite different, with rich chestnut on the back and a black patch beneath.

Juvenile Ruffs

A party of ten juvenile Ruffs stand close together in still water that reflects them perfectly. The diamond pattern of the group becomes a hexagon. Most of the birds which pass through Britain on migration in autumn are young ones and, to judge from the uniformly buff colour of their breasts and by the leg colour, these are all juveniles. It is hardly possible to say with certainty whether they are Ruffs or, as the smaller females are called, Reeves, though there is a tendency towards sexual separation in winter and, from the uniformity of size, these are perhaps all one or the other. This painting is thought to have been exhibited at the Royal Academy.

Curlews Alighting

An autumn flock of Curlews is coming in to land on a strip of sedge marsh with the estuary and the distant hills beyond. It is a fine day of blue sky and piled cumulus clouds and the composition closely resembles that of the picture of Redshanks alighting, though here, with no water in the foreground, reflections play no part. There is the same shrewd observation of the sequence of movements in descending from level flight to a landing on the support of long, brittle legs and similar play with the sharp contrast of sunshine and shadow on the birds.

Redshanks Alighting

Here we have a flock of nine Redshanks alighting on an area of mirror-still pools on a sandy beach. Resting on the sands there is a distant flock of gulls visible as little more than an elongated patch of white but, nevertheless, skilfully suggested so that we may safely guess that they are Herring Gulls with one or two Great Black-backed Gulls among them. The Redshank is sometimes known as the 'warden of the marshes' because it is usually the first bird to notice an intruder and to rise with its loud, melodious call which effectively draws the attention of other birds to the danger. In spite of its bright red legs, when feeding on marsh or shore the Redshank is not a conspicuous bird, but when in flight, besides the insistent cry, the bold white rump and the broad bands of white on the rear of the wings make it very apparent. Being so common about his home this bird constantly recurs in Tunnicliffe's sketchbooks and there is at least one carefully measured drawing in his collection.

Lapwing Family

This oil painting is one of several of Tunnicliffe's pictures of a breeding pair of Lapwings with their four newly-hatched downy young. One, the male, with spread wings, has just alighted. The female has a shorter crest and more white on the face. The eggs have only recently hatched in a nest on drier ground a short distance away and the young have been led to the edge of the pool where food is easier to find. If alarmed the downy chicks will crouch motionless, their white patches entirely concealed and the little birds all but invisible while the parents do their best, by noise and distraction displays, to deflect the intruder's attention.

Lapwing Flock

The Lapwings here, unlike the pair opposite, are not in breeding plumage. It is winter. The birds have gathered in a flock, pairs and territories forgotten. They have much buff colour about the head and the white on the upper breast intrudes on the black breast-band. They are resting and preening on the tussocks of grass along the irregular edge of the Cob Pool. They all face to the right, into the breeze no doubt, as birds do when resting. One, preening the upper tail coverts, has its tail fanned to display the bold black arc of the subterminal band. This pose breaks the regularity of the group and the tail acts as a foil to the curves of the grass verge to the right of the picture.

Black-tailed Godwits

Black-tailed Godwits in the handsome red spring plumage usually appear in Anglesey in April and, though they do not stay to breed, they often visit the Cob Pool for a few weeks. This picture shows a pair resting in shallow water with a background adorned by water crowfoot in bloom. The design has a strongly diagonal form with the lie of the crowfoot and the foreground mud spits serving to echo the angle of the birds' bodies and bills. The necessary verticals of the godwits' legs and their reflections are tempered by clumps of sword-like young reeds in the background.

Bar-tailed Godwits and Dunlin

The larger birds are Bar-tailed Godwits, close relatives of the Black-tailed Godwits opposite. The small birds are Dunlin. Like Black-tailed Godwits, the Bar-tailed Godwits have a red breeding plumage but it is rarely seen in Britain because their breeding area is in the arctic, whereas Black-tailed Godwits breed in Holland and, increasingly, in parts of Britain. Bar-tails appear on the Cefni estuary in late summer and are more given to open beaches, as depicted, than the Black-tails, which prefer mud and fresh water to sand and sea. Two of the Dunlin are still in breeding plumage, with black bellies, the others are juveniles. The composition is a skilful elaboration of a diamond-in-rectangle pattern.

Night Herons

The Night Heron is a very widespread species in the world but in Europe is confined to the warmer, southern parts and is no more than a scarce and accidental visitor to Britain. It is common in parts of North America, where it is distinguished as the Black-crowned Night Heron from another species, the Yellow-crowned Night Heron. It is largely crepuscular in habits and in the daytime roosts in dense vegetation. I suppose this elegant vertical composition to have been based on studies of captive birds.

Grey Herons

Tunnicliffe called this picture of two sleeping Herons 'Heron Cove', suggesting that he knew of some pleasant nook in the rocks that Herons frequented when resting. However, the birds in this pose were in fact sketched whilst sheltering between a wall and a heap of bricks! They are perched on an irregular ledge just above high tide mark where the rock is adorned with dark patches of seaweed. This channelled wrack is less dependent than most on regular sea-water soaking and consequently grows at the highest inter-tidal levels. The dark areas of weed are skilfully disposed to act as a foil to the black patches on the birds and the blue-grey of the upper shadowed rock extends the preponderant grey of the herons.

GULLS AND TERNS

The graceful beauty of terns – 'sea swallows' -- was a source of great pleasure to Tunnicliffe. Apart from a small group of so-called 'marsh' terns, of which the Black Tern made an occasional appearance on passage at the Cheshire meres, terns are essentially sea birds. They are scarce in Cheshire but Tunnicliffe was well placed to study them at Malltraeth. There are, or were, large breeding colonies in the immediate neighbourhood consisting principally of Common and Arctic Terns usually with a few pairs of the relatively rare and lovely Roseate Tern. Little Terns breed, but not colonially, along the beaches and are often seen fishing over the Cefni estuary and the Cob Pool. The Sandwich Tern also breeds in Anglesey but is less often seen than the other species in the vicinity of Tunnicliffe's home at Malltraeth. All these terns are summer visitors and, to those living on the shore as the artist did, were as welcome as swallows when they arrived in April and May. Resting flocks of terns adorn the beaches throughout the Anglesey summer and make lovely patterns in their setting of water-rippled sand or of sand and sand-pools. In a breeze they crouch low, head to wind, their delicately streamlined forms in themselves a picture of grace and beauty. As soon as they arrive they engage in delightful nuptial displays with wings held low and angled away from the body, necks stretched high and tails pointed upwards, giving rise to strange angular forms. In this posturing they run round each other, sometimes in groups of several birds, and make the most attractive of patterns. In *Shorelands Summer Diary* Tunnicliffe recalls the afternoon of 29th June, when he and some friends were watching Arctic terns:

We had tea in the lee of the pebble ridge and watched the lovely flight of terns. Their elegant forms reflected the light from the sunlit shingle and appeared almost incandescent against the blue of the sky. Two of them performed the daintiest of aerial ballets, swooping up together almost vertically, to meet at the top of their swoop and to touch bills. Again and again this flight was repeated and in it there was more of the quality of butterflies than of birds, for wings and tails were spread to the full at the height of the flight. One tern hovered and circled overhead and seemed very uneasy at our presence. Soon it dropped to the shingle, about forty feet away, and shuffled down as if brooding eggs. Here was a nest I had missed. She sat with her head and neck and the wing tips and tail streamers showing above the large pebbles. Any sudden movement by us put her into the air again but she soon returned, making an indescribably beautiful landing. I focused the glasses on her and, to my surprise, saw that her blood-red bill had no vestige of black on it. When she first flew we had commented on the length of her tail streamers and, as she alighted on the pebbles, the shortness of her legs. The quality of the red of her bill was different from that of the Common Tern's and we had no doubt that

Common Terns

A group of Common Terns rests on a rock by a pool on the shore, heads to the prevailing wind which veers in gusts from side to side. The feathers of the birds are intermittently lifted in the breeze as they crouch lower to present the smallest surface to its discomforting effect. The water of the pool is churned into ripples and wavelets. Here Tunnicliffe exhibits not only a sound knowledge of what terns look like, but also displays the fruits of careful observation and skill in execution in conveying a convincing impression of a particular kind of day, even to the line of spume along the tide-edge, blown, in places, away on the wind.

she was of the Arctic species. In her "nest" were two eggs of the same large blotched dark tint as in the one I had found. After tea the others left me to draw while they went in search of Little Terns' nests.

Gulls, of course, are present all the year round on Anglesey shores, although the species do vary a bit with the season. The most constant at Malltraeth were the Black-headed Gulls and Charles painted pictures of them many times. I remember him saying, during one of our later meetings, that he 'hadn't finished with Black-headed Gulls yet'. He was particularly pleased with the picture he called 'The Ninth Wave'. Here he strove, very successfully, to convey the almost fairy-lightness of the birds as they allowed the wind to lift them just sufficiently for the big wave to pass beneath; before alighting to resume feeding in the next trough. It was characteristic of him that, while not insensitive to the excitement of a passing rarity, he was happy to continue studying the common birds about Malltraeth, for he was primarily interested in the decorative possibilities of the bird in their various surroundings.

Gulls are one of the least popularly appreciated groups of birds especially as they have, from being just the 'seagulls' that lend atmosphere to a seaside holiday, become the noisy nuisances that frequent city rubbish tips and are alleged to pollute reservoirs. Even to bird watchers they were usually, until recently, dismissed as just 'gulls' and only rarely examined closely. But habits and fashions change and in recent years gulls have received more detailed attention and much has been learned about them. Many of the larger gulls take several years to attain the grey and white 'seagull' plumage and pass through baffling annual stages of varying brown and white. These immature brown plumages were at one time generally regarded as impossibly complicated but recent studies have led to greater clarification and understanding. Tunnicliffe, however, always paid great attention to the juvenile stages, finding the delicate pencillings in the plumage of the young birds a special fascination. I remember a picture that he called 'Sea Lace' in which he contrasted the dark-on-light pattern of two flying juvenile Herring Gulls with the light-on-dark network of a wave running down from a rock.

Sunlight and Shadow

This is an oil painting. A resting group of Common Terns is being joined by two more. The sands are golden and the terns are reflected in a still blue sand pool. In the distance the sea rolls gently onto the shore where a flock of Herring Gulls and three Oystercatchers are resting. The horizontal composition is relieved by the raised wings of the alighting terns. Sunshine has produced strong shadows. The dark blue water and rich golden sand, with the white and pearly grey terns showing crystal bright, is a common and entrancing spectacle on the Anglesey coast in spring. Tunnicliffe in his maturity rarely painted in oils unless specifically commissioned to do so, but he was equally competent in this medium as in his more practised water-colour.

Gull Gallery

Two Great Black-backed Gulls are surrounded by a group of twelve Common Gulls and the party rests, heads to the breeze, on a rock by the shore, one or two of the birds idly preening. The Common Gull is common only in winter in England and Wales and it may be that the picture recalls a group seen by Tunnicliffe when on holiday in Western Scotland, where it breeds in numbers. The birds are evidently in breeding plumage with heads unsullied by grey streaks and their bills brightly yellow. The Common gull is expanding its breeding range southward and there are already a few colonies in Anglesey. Moreover, since Scottish birds begin to move southwards even as early as July, it is possible that the picture was in fact based on studies made close to the artist's home.

Ringed Plovers

These Ringed Plovers are resting on shiny wet sand that reflects a blue sky. They are grouped diagonally, two of them breaking the formation with raised, stretching wings. A few, with incomplete collars and less boldly marked heads, are juveniles, birds of the year. The expanse of sand round the birds is broken by several strands of thongweed cast up by storms from the deeper zones of the shore. The serpentine fronds of the weed behind and the stipe and holdfast in the foreground convert the diagonal of the birds into an S-pattern; a form of design the artist often favoured.

Black-headed Gulls

After the breeding season is over Black-headed Gulls lose the dark brown hood, on which their name is based, and there remain only a few dark smudges on the white head. Two of these birds are young ones, birds of the year, as can be seen by the brown feathers in the wings and by the bill colour, yellowish with a dark tip. In the breeding season these are a dull, dark purplish red that fades to scarlet as the year advances.

Wader Reflections

Four Oystercatchers, three Redshanks, four Ringed Plovers and five Dunlin are gathered together on a patch of wet sand that reflects them and the blue sky to produce a slightly smudged inverted image. It is probably late summer. The birds are no longer in breeding pairs but they have not yet moulted into non-breeding plumage. The Oystercatchers lack the white half-collar of winter and the Dunlin still have the black belly of summer plumage. Such highly decorative mixed groups of waders were a constant source of interest to Tunnicliffe and he made many pictures based on this general theme, often, as here, with the boldly black, white and red Oystercatchers as a focal point. This picture is included in the 'Gulls and Terns' section because it makes a pleasing pair with the one opposite.

Sandwich Terns

A group of six Sandwich Terns rests on a muddy shore on a still day among shallow pools left by the tide in the hollows of the wave-pattern on the beach. There are lugworms in their U-shaped burrows deep in the mud. This is evident from the little convoluted heaps of mud that decorate the shore. These are the excretions of mud that has been passed through the worm's digestive system and squirted to the surface. One or two cockle shells may also be seen. That the terns are juveniles is evident from the blackish bars and spots on the feathers of their backs and wing coverts.

Black-headed Gulls

Tunnicliffe's title for this picture was 'The Ninth Wave'. The ninth wave is, according to him, reputedly the largest wave. More usually it is said to be the seventh. A party of Black-headed Gulls has been feeding in the shallows by the shore, rising and falling with the lesser waves. As the big wave rolls in, threatening to overwhelm them, they lift gracefully into the air and let it break beneath them, dropping again to feed in the trough when it has passed by. It is winter and the gulls lack the black head of the breeding season. A few are birds of the year with black bands in the tail and a brown mottling on the upper wings.

AUKS

We reproduce here one of the paintings that Charles Tunnicliffe made of Puffins resting on their flowery clifftops in spring. This is a very special sight for the bird-watcher but it is not easily available. Puffins nest on remote and, usually, uninhabited islands. Charles made at least one visit to Skokholm and Skomer islands, off the Pembrokeshire coast, where hundreds of Puffins may be seen in spring and early summer in the sunshine, in groups on the grassy turf in which their burrows are excavated, or on the neighbouring rocks. At that season the clifftops are veritable flower gardens with massed blooms of pink thrift, white sea campion and yellow kidney vetch. The rocks themselves are beautifully decorated with orange, grey and green lichens. Puffins are, if approached with reasonable caution, very tame indeed and can be drawn from life at close quarters. They seem more easily disturbed if approached on foot but by a belly-crawl one can study them almost literally face to face without difficulty. They are irresistibly and comically, attractive in their portly demeanour and, though feathered soberly in black and white, have multi-coloured bills, in which scarlet predominates, and bright orange legs and feet which seem designed to harmonise with sea-pinks and orange lichen.

At South Stack in Anglesey the artist had ready access to mainland cliffs where other species of auk, Razorbills and Guillemots, nest in large numbers, but these relatively soberly clad birds seem to have made a lesser impact on him. In '*Shorelands Summer Diary*' they are readily deserted for the greater attractions of the Peregrine's eyrie and I do not remember a picture that he painted of them, though doubtless he did not ignore them entirely.

Puffin Colony

This is one of a number of pictures made by Tunnicliffe of a group of Puffins resting among spring flowers on a cliff top. The colour and formation of the rock suggests that it is based on sketches made on a visit to Skockholm in Pembrokeshire. There are a few Puffins to be found breeding in Anglesey and Charles had sketches made at Gull island off Aberdarn at the top of the Lleyn Peninsula, but they are much more numerous and easy to watch on the Pembrokeshire islands. Puffin island, off the south-east coast of Anglesey, is no longer aptly named. The Puffin colony is much reduced.

BIRDS OF PREY

From the time when, required by Henry Williamson to illustrate '*The Peregrine's Saga*', Tunnicliffe went to falconers meets and visited individual falconers in order to study birds of prey, he had a deep affection for these fierce hunters. In '*My Country Book*' he describes with manifest enthusiasm a falconers' meet that he had attended at Avebury in Wiltshire. But it was when he went to live in Anglesey that he was able to virtually live among birds of prey. The estuary before his window, with its gulls, wildfowl and waders, attracted Peregrines, Merlins and even the occasional Gyrfalcon in winter. Sparrowhawks would appear in pursuit of the small birds in his garden. Best of all, on an outing to study auks at South Stack, a steep and spectacular headland near Holyhead, he found a Peregrine's nest on the cliffs there. As he describes in his '*Shorelands Summer Diary*' (Collins, 1952), he visited this eyrie at regular intervals during that summer and studied and sketched the birds at every stage from his earliest observations of the adults brooding tiny chicks to the final wing exercises of the fledged eyesses before they ultimately left the neighbourhood. The picture of a tiercel that we reproduce is one of the splendid paintings that emerged from this glorious summer with the falcons at South Stack.

Living in Wales is, of course, to live among Buzzards, and Charles had many sketches and measured drawings of these birds and would see them regularly on his outings. The picture he called 'Buzzard in the Rain' is one of his most successful compositions.

The roving and unscrupulous man with a gun, and the official keepers of game, were both apt to destroy any bird with a hooked bill and sharp claws and a number of specimens found their way to Tunnicliffe's studio to enlarge his collection of measured drawings: several Peregrines, Kestrels, Sparrowhawks, Buzzards, Merlins, Owls (Barn, Little, Long-eared and Short-eared) and even the winter rarities, Gyrfalcons and Snowy Owls.

Although ornithologically classifed remotely from the hawks and falcons, owls are the birds of prey of the night and Tunnicliffe's affection for them extended back to the Cheshire days when he used to study them on Goldsitch Moss, an area of bleak moorland not far away from the farm. He was always excited by the spectacle of Short-eared Owls hunting the moors, as he described with enthusiasm in '*My Country Book*'. He was also in Short-eared Owl territory in Anglesey for they hunt in winter over the Cefni marshes and over Newborough Warren. Newborough Warren is a partly forested sand dune wilderness area adjoining the Cefni estuary. It is one of the many warrens in the British Isles where 18th and 19th century landowners maintained stocks of rabbits for the table, in the days when fresh meat could not be preserved for very long. It remains a natural hunting ground for birds of prey. The painting here of a Short-eared Owl is a somewhat unusual picture for Tunnicliffe; it depicts the flying bird in full detail, far more

Buzzard

A Buzzard rests on the branch of an ash tree on the banks of the River Tywi on the Welsh mainland. Across the river on a knoll are the ruins of Dryslwyn Castle. In the meadows below the castle a herd of Hereford cattle graze. It is typical Buzzard country where the hillsides are patched with hanging oakwoods. Buzzards, unlike many species of birds of prey, perch in conspicuous and exposed places and this circumstance has enabled Tunnicliffe to make a convincing picture that is part bird study and part landscape without losing any reality of scale, as so often happens when an attempt is made to relate a bird to an extensive view. In the Middle Ages, Dryslwyn Castle played an important part in the struggles between Welsh and English.

detail than could ever be discerned. Usually he contended that a picture should represent what can be seen and should ignore detail that an observer would be unable to perceive in real life with binoculars. It is also not a typical Tunnicliffe painting in its almost total disregard of composition.

Occasionally the wandering Gyrfalcons that visited Anglesey in winter were white ones from Greenland. When one of these was found, shot but still alive, on the banks of the Cefni river, Tunnicliffe was able to make sketches of it before it died. Unfortunately he could not make a measured drawing of the corpse before it was sent to a taxidermist. No doubt these sketches were an important basis for his paintings of White Gyrfalcons, one of which we reproduce. The Snowy Owl picture was derived from a bird shot at Mynchady but in that case Tunnicliffe had been able to make a detailed measured drawing of the body.

Short-eared Owl

This picture shows a Short-eared Owl hunting over heathy land near the sea. Darkness and a stormy sky are imminent but there is light enough to see the bird clearly. Short-eared Owls are crepuscular rather than nocturnal and often hunt the marshes about the Cefni River and the Cob Pool in broad daylight. The British population is greatly augmented in autumn by immigration from the Continent. They appear at the time that the Woodcock come and are known to some as Woodcock Owls. Another arrival at that time, the Goldcrest, the Woodcock's Mate, was at one time supposed to ride on the Woodcock's back.

Peregrine

This is a splendid portrait of a Peregrine. Tunnicliffe called it 'Tiercel Perch'. The bird is an adult male, or tiercel, as male falcons are known, the word 'falcon' being reserved, strictly speaking, for the female. The term 'Tiercel' is said to originate from the circumstance that in most raptors the male is about one third smaller than the female. Tunnicliffe had a special interest in birds of prey and, of these, more particularly the Peregrine. Although the picture is a simple portrait in a panel format with the bird in a vertical posture, the composition is ingenious and satisfying in a linear sense with a diagonal emphasis in the setting as a foil to the verticals of the figure. There is also a skilful and harmonious arrangement of colour and tone: the white breast of the bird echoed in the foaming water, the grey shapes of the upper parts repeated in the grey sea and the bright yellow of cere, orbital ring, legs and feet incorporated in a pattern of yellow with that of the flowers on the clump of hawkbit. The green of the leaves of the plant is continued in the patches of lichen on the sunlit cliff.

A Buzzard is perched on a shattered and rotting tree stump. Grasses grow on the broken upper ends where leaf-mould has accumulated. The fruiting bodies of honey fungus are growing lower down. It rains. In the blue distance the shape of a conifer is just visible. Like the plate opposite, this is a simple portrait of a bird of prey in panel format made into a delicately exquisite picture by immaculate design. The Buzzard's tail is echoed in the crotch of the tree, the scalloped back feathers of the bird are repeated in the structure of the fungus and grass formations provide subdued echoes of the Buzzard's head.

Snowy Owl

When Tunnicliffe made a life-sized measured drawing of a dead Snowy Owl he noted that it had been shot at Mynachdy in the far north-west of Anglesey, just inland from Carmel Head. This picture is an imaginative composition showing what Tunnicliffe conceived the bird might have looked like before it was shot. The view is of Carmel Head and the island with a lighthouse out to sea in the squall is the West Mouse. The bird is a female, as is the specimen he drew in the studio. The male is more nearly an immaculate white. The Snowy Owl is an arctic species that moves south in winter and appears in small numbers in Britain and along the Eastern Seaboard of the United States. For a few years a pair bred on Fetlar in the Shetlands but, latterly, the female has appeared to lay sterile eggs.

Gyrfalcon

This painting of a Gyrfalcon probably represents a bird from North Greenland where many birds of this species are white, or nearly so. Those from Iceland or Norway are rarely so white and indeed many are as dark as Peregrines. White Gyrfalcons are relatively common in winter in North America but they are very rare visitors to Britain. Gyrfalcons indeed occasionally appear in the Cefni estuary and were seen from Tunnicliffe's studio window but so far as I know all these were grey birds and frequently juveniles. However Tunnicliffe did have sketches of an injured white bird that no doubt facilitated this imaginative picture. Perched on a lichen-patchy wall the bird looks out over, possibly, the Menai Straits, towards the snow-clad hills of the Welsh mainland. The composition, as in the picture of the Snowy Owl opposite, is firmly pyramidal.

Goshawk

Here we have a splendid picture of a Goshawk. It has caught a bird, not easily seen but possibly a Partridge, and is 'mantling' over its prey in deep drifted snow from which a thorn bush protrudes elegantly to aid the rotary design. The Goshawk is a bird of woodland and is unlikely to have been seen in Anglesey, but Tunnicliffe knew the species well from his studies of birds of prey at falconers' meets and in the collections of falconers. Formerly extinct in Britain it is now again present in some numbers, perhaps partly by colonisation from the Continent but also, doubtless, from escapes in the course of falconry.

Lanner

Tunnicliffe's interest in falconry and its appurtenances is displayed here in a portrait of a trained falcon. In addition to the careful and detailed presentation of the bird Tunnicliffe has shown with equal care the 'hood', with its decorative crest and straps at the nape for instant release, the 'jesses' short straps attached to each leg and bearing bells, the 'leash' to which the jesses are attached by a swivel, and the 'block' on which the falcon is being 'weathered'. All are painted with great care and attention to detail and there is little or no concern with design in the faithful representation of things as they are. The painting of the wooden block is worthy of note as is the meticulous treatment of the grass, doubtless achieved by use of masking fluid. The falcon in the picture is not a bird likely to be seen in Britain in the wild. It is a Lanner, essentially an African bird which is not uncommon in the far north of that continent, but is very scarce in Europe. It is perhaps best distinguished in the field from the Peregrine by the pale buff crown and nape but that feature is not visible in this hooded bird.

FOWL AND GAME

Tunnicliffe's affection for, and interest in, the creatures of his childhood farmyard never left him, and he continued to make fine pictures of the Cocks and Hens and the Turkeys. In Wales, still a land of the small mixed farm, barndoor or even kitchen-door poultry could still be seen at a time when in England they had all but disappeared into batteries and the 'cock's shrill clarion', once one of the most nostalgic of village sounds, was heard no more. Some of Tunnicliffe's finer paintings are of Cockerels blown by the wind and of poultry and Turkeys roosting in trees. These sights are available only where fowl are 'free range' as we now must call them, to distinguish these oddities from the tightly boxed denizens of egg factories that are poultry today.

The picture of poultry roosting in a damson tree is evidently derived from a very early experience, for he says in '*My Country Book*':

> *One evening in spring I saw an exquisite sight. A white Leghorn cock and four hens had gone to roost in a damson tree which was in full bloom. The combination of white, brown and chequered fowl and starry white bloom on thin black twigs, backed by a delicate spring evening sky, was a joy to behold.*

Turkeys, even more so than poultry, have almost completely disappeared from the farmyard and it is a pity because a Turkey Cock in full display used to be one of the more splendid if alarming spectacles to be seen in an English village. Tunnicliffe greatly admired these glossy inflated birds with their self-important gobbling and apoplectic faces. In the picture reproduced of a wholly white one in a winter setting, some discarded and rusty cartwheels, frosted over on an icy morning, are used as a counterpoint to the white fan of the Turkey's tail. In the picture that he called 'Family Tree' a family of normally coloured Turkeys are roosting in an ash tree.

The picture of Game-Cocks in battle represents another facet of Charles Tunnicliffe's lively interest in animals and birds. Besides being a splendid design in flowing curves the painting is a hymn to the striving male birds in the full magnificence and pride of their lusty prime. His fondness for farm animals was usually focused on the combative male, especially the dangerous farm bull and the 'majesty at ease, the fruitful pride' of stallions which, led by their 'buskined grooms' used to parade the countryside on their way to their duties. I remember him once saying how much he would have liked to go to Spain. Not to lie festering and idle in the sun but to see a bullfight in all its gorgeous pageantry and bloody conflict – and paint it.

Pheasants, although introduced birds (by the Normans, it is said) are, aided by game preservation, common and highly decorative inhabitants of the countryside wherever there is sufficient cover for them. Tunnicliffe was gripped by the splendour of the Cock Pheasant with his purple and green glossed, black crested,

Fowl in a Damson Tree

A group of poultry roost in a damson tree in bloom. This is one of Tunnicliffe's most Japanese-style pictures with an irresistible flavour of the oriental. The birds are probably Leghorns, the cock a particularly splendid specimen. They are lit from below, as they might be by a torch, as the farmer searches for birds roosting hazardously in the open when they should be in a fox-proof cover. Even the dark-brown hens above are light against the dark night sky and the blossom shows up as white as in daylight. Tunnicliffe's title was, simply, 'The Damson Tree'.

head and bare scarlet face, shiny chestnut red black-laced plumage and the long darkly barred tail. He painted many pictures in which these magnificent birds featured. Cock Pheasants fighting, Cock Pheasants displaying to the dowdy female, Pheasants just being Pheasants. He often placed them in a setting of autumn-golden bracken the pattern of which harmonises so well with that of the bird's plumage.

At a place called Trefeilir, near to Tunnicliffe's home in Anglesey, a breeder kept a collection of highly decorative ornamental pheasants. There were Golden, Silver, Elliot's, Lady Amherst's, Reeve's and Eared Pheasants, among others, at Trefeilir, and these fine birds of oriental origin moved him to eastern forms of design. He admired the Chinese and Japanese artist's talent for decorative composition and their influence is apparent in his pictures of this group of birds; especially so in the vertical panels, of which the one of Blue-Eared Pheasants is particularly successful.

Closely related to the pheasants are the Peafowl and at Trefri, a farm also near to home, Tunnicliffe was able to study them at leisure. They are commonly kept ornamental fowl with qualities as watch-dogs that are sometimes appreciated. In '*Shorelands Summer Diary*' Charles tells of his first visit to Trefri and of his surprise at finding green lawns and Peafowl on that rocky and rabbity wild and remote headland. He remarks on 'a certain elegant white Peacock, which I wanted to allude to as "she", so delicate did he look among his more gorgeous companions'. Two of his pictures of Peafowl are reproduced here, a water-colour panel of the birds in a tree and an oil painting of a group resting on the ground.

Partridges are, or were, numerous in Anglesey, but only the native 'grey' Partridge. The introduced and flourishing Red-legged Partridge that is more common than the native now in some parts of England is not found there. The appreciation of elegance in shape that drew Tunnicliffe to Peafowl and Pintails did not keep him from making pictures of Partridges, even in winter time when their always dumpy forms become almost spherical in the cold. Seen from a distance both species appear as brown blobs in the field but at close quarters their delicately pencilled feathering is most beautiful. He found either species very lovely in a setting of equally delicate winter plants picked out in white fronds by hoar frost.

Cock in the Wind

A handsome cockerel contends with a vigorous wind that has taken him in the rear and turned all his hackles and plumes the wrong way round, all but toppling him over, head first. Besides being a fine study of a particularly splendid cock there is in this picture an element of humour in the discomfiture of so much masculine pride. As in the painting of the white cock (page 109) the impression of wind and movement is augmented by the heeling grass and swirling dead leaves which at the same time serve to complete the design.

Sparring Cockerels

These are Game-cocks, of the kind formerly bred for fighting but now produced only for show purposes. They are lean and muscular creatures with comb and wattles trimmed to provide no grip for an enemy. Spurs were formerly replaced by sharpened steel. Tunnicliffe had a number of sketches of Game-fowl from birds kept by a friend. His composition gives a fine impression of the swirl and scuffle, thrust and parry of cockerels in conflict. The birds all but fill the picture rectangle but studies of dandelion rosettes echo the fanned wings and sprays of dwarf mallow accentuate the rotary composition.

Ring-necked Pheasants

Two handsome cock Pheasants with fiercely fanned tails dispute possession of the dowdy hen in the background. Tunnicliffe called the picture, oddly, 'Faisan Triangle'. Unlike the game-cocks in the picture opposite, these birds are unlikely to do much more than seek to intimidate each other by a threatening display of gaudy plumage. One of them, after a few passes, will probably compact its feathers and slink away. Pheasants are introduced birds in Britain and the population is perhaps sustained only by the game preservation, breeding and release programmes of the shooting fraternity. The combatants in the picture are of the ring-necked variety, originally from China, which has tended to displace the original introductions of darker, ring-less birds from the Caucasus.

Blue-Eared Pheasants

Here Tunnicliffe has made a very pleasing design in panel format of a pair of Blue-Eared Pheasants perched in a rhododendron, the composition strongly suggestive of Japanese influence. Eared pheasants are remarkable in the feathered 'ear' tufts and in the magnificent tails with the central feathers elegantly curved and with highly disintegrated webs. They are unique among the pheasant tribe in that male and female are barely distinguishable. Commonly among pheasants the hen is in dowdy brown contrast to the splendidly colourful cock. The Blue-Eared Pheasant is a very local bird of mountains in Western China, known in Britain only as a captive specimen in zoos and in private collections.

Lady Amherst's Pheasants

This is another design of oriental style based on a pair of Lady Amherst's Pheasants. They are roosting in some kind of conifer which has laxly drooping needles as a convenient echo of the long pointed and barred tails of the birds. Lady Amherst's is one of the two so-called ruffed pheasants, the other being the Golden Pheasant. Very popular in collections of exotic pheasants, both species have been released locally in Britain. They are disliked by sportsmen, being reluctant to fly, and although well-established in certain localities, neither has spread extensively.

Grey Partridges

The common or grey Partridge is the subject of this picture. A covey of seven chubby birds lurks in the early morning among hoar-frosted vegetation. The mists of night have not yet quite cleared and the delicate forms of the grass picked out by frost provide an echo of the fine barrings and pencillings on the birds. The Partridge is much less common that it used to be and, being a favourite game bird, the causes have been much studied. It appears that Partridge chicks need an insect diet which they formerly found among grassy field headlands and by hedgerows. More thorough and mechanical farming have almost eliminated these narrow strips of uncultivated land. Insecticidal sprays have done the rest.

French Partridges

The painting above of four Red-legged Partridges is another in which Charles Tunnicliffe made use of hoar-frosted vegetation. The Red-legged Partridge, or 'French Partridge', is not a native of Britain but was introduced from the Continent during the last century. In this picture the four birds are fluffed out so that their normally portly forms are all but spherical. The blue shadows on the frosted vegetation are repeated in the blue of the birds underparts and there is a precise echo of the white patches on the faces of the partridges in the shapes of (possibly dock) leaves in the foreground. The curves of the frosted grass to the left of and behind the birds provides a foil for the hunched backs of the crouching figures. The net-like pattern of white twigs against a dark trunk is repeated, reversed, in the dark shadows of the ruffled back feathers. The frosted oak leaves are curiously, though perhaps hardly necessarily, reminiscent of the bold flank patterns of the birds.

Roosting Turkeys

Turkeys are difficult to rear under farmyard conditions. The chicks suffer from damp and disease and die off alarmingly. But before the development of factory-like poultry production, barndoor Turkeys were to be found with the chickens on a few farms. Sometimes, where foxes are scarce, they were allowed to roost out of doors. Tunnicliffe has made a fine picture, albeit almost monochrome, from a Turkey family roosting in an ash tree where the dark pattern of the ash leaves against the night sky provides an echo, in reverse, of the pale tracery of the pattern of feathers on the birds. The wattled red and blue of the Turkey cock has introduced a small accent of subdued colour. Tunnicliffe called this picture 'Family Tree'.

Black Grouse

Black Grouse are common birds in the pine woods of Scotland and there are small numbers in England and in Wales. Tunnicliffe had measured drawings of specimens sent to him from East Cheshire. Blackcock (the male, the female is known as the Greyhen) are most spectacular when seen at the display ground, known as a 'lek'. Here the Blackcock gather and seek competitively to impress the Greyhens by inflated neck, drooped wings, spread tail, much 'rookooing' as the call is known and some ritualised fighting. In this picture Tunnicliffe has presented the birds in the depth of winter after heavy snow. A group of three Blackcocks and two Greyhens is resting on a pine branch. The buds and needles of the Scots pine are a staple winter food of the Black Grouse.

Peacock and Consorts

This picture, like that opposite, is a study of a Peacock with two Peahens, but in this case Tunnicliffe has adopted a horizontal format and the birds are portrayed squatting on the ground. In what is perhaps a more sophisticated and original design the Peacock's intensely blue head and neck are the focus, with the vertical balanced by the neck of the foreground Peahen. The reclining Peahen behind the cock has the neck curved back and, instead of providing a third vertical emphasis on the heads, takes interest to the balancing curve of the Peacock's tail. The white areas are nicely adjusted, as are the delicately patterned grey-brown of the foreground Peahen and that of the cock's wing coverts. Little fans of orange grass provide an echo of the fan-like head decoration of the birds and of the dominant colour in the tail 'eye' of the Peacock. This painting was carried out in oils.

Peafowl Roosting

Peafowl are highly decorative birds that have been kept in parks and pleasaunces for many years. They would probably have been even more popular were it not for their loud and raucous cries. Tunnicliffe had access to several places in North Wales where he could study them. This picture shows a Peacock with two Peahens perched in a coniferous tree whose pine needle clusters repeat the ornamental 'eyes' of the Peacock's tail, and an echo of the fans of crown decoration on the heads of both sexes.

White Turkey

A white Turkey cock is in the course of his magnificent display. His tail is spread into a perfect fan, the bare bulbous skin of his head and neck are engorged and scarlet. The wings are spread and scrape the ground as he moves jerkily forward uttering his gobbling call. It is winter and the rough herbage of the farmyard corner is white with rime. The Turkey, alone, is a splendid picture and the setting is designed to augment his pride. The white spread of his tail is set off by the frost-encrusted discarded cartwheels and the spread of his wings is echoed by the rigidly vertical spikes of frosted dock and the dimly perceived fence behind him. The inflated scarlet of his head is not quite the only warm colour amid the blue and white chill – there is a suggestion of orange-brown in the wheels and in the dock leaves. The artist called this picture 'Winter Turkey'.

Leghorn Cockerel

This painting illustrates Tunnicliffe's interest in movement and his skill in conveying it in his pictures. A magnificent White Leghorn cockerel is caught in a gust of wind and is barely able to maintain his stance. The wind blows his tail sideways and inverts his neck feathers into a ruff. He turns his head from the breeze and his left foot is lifted as he all but loses his balance, his lee wing spread as a stay. The leaves of the mallow at the bottom left are turned inside out by the gust as dead leaves swirl and dance on their way. A masterpiece of careful observation that leaves the observer holding firmly to his hat.

Reeve's Pheasants

This picture and the one opposite represent birds from the groups of so-called 'long-tailed pheasants'. Here we have a panel composition, in an oriental style, showing Reeve's Pheasants perched on the branches of a flowering tree, probably a plum tree. There are two males and one female. The white plum blossom continues into the background the white-spotted pattern of the cock birds. Reeve's pheasants indeed have very long tails. They are popular with collectors and have been released into the wild from time to time, with only temporary success. Their native home is Western China.

Elliot's Pheasants

Elliot's, another species of the 'long-tailed pheasants' from China, has a tail not nearly as long as that of the Reeve's. The design, showing one male and one female perched in a maple tree, is very similar to that opposite and is markedly eastern in flavour. Elliot's Pheasant does not survive for long if released unless the eggs are gathered and the chicks reared in captivity under broody hens. Consequently the species is less well-known generally than are Golden and Silver Pheasants (and to a lesser extent Reeve's Pheasant) all of which are successful park birds and could become established.

PIGEONS

Tunnicliffe's collection of sketchbooks include several that were devoted entirely to studies of fancy pigeons. These were made at the aviaries of a breeder of pigeons in Anglesey and were carried out in order to make paintings of each of the many breeds and varieties for the owner himself. As a result of these studies he also painted a number of exhibition pictures, of which the magnificent one opposite is a fine example.

When making a number of pictures from one species of bird, Charles thought it would be tedious to distinguish them for catalogue purposes as, shall we say, 'Mallard 1', and 'Mallard 2'. Instead he invented distinctive titles which were often witty or whimsical and indeed sometimes a little obscure. Where a title for a picture is known I have mentioned it, for example, 'The Ninth Wave', and 'Heron Cove'. His title for the picture opposite was 'Angels and Archangels' but I can give no explanation for this. The number of varieties of pigeon that have been bred is legion and the names given to them are, to anyone unfamiliar with the jargon of the fancier's trade, almost fantastically diverse and exotic. This is apparent from the sketchbooks, which include drawings of The Red Lodz Tumbler, The Black Pygmy Pouter, the Viennese Gauzel, the Dun Mookee and the Spangled Priest, but there is no mention of either angels or archangels. However experts say there *is* a pigeon called an 'archangel' although apparently it is not represented in this picture. The birds, I think, are two Jacobins and a Fantail. The flowers in the background are white aquilegias, or columbines, not known, so far as I am aware, either as angels or archangels. Perhaps the painting, when completed, put Charles in mind of his conception of a heavenly host.

Though a great deal of time must have been devoted to the studies of fancy pigeons in his sketchbooks it would be true to say that, apart from this exercise, the pigeon family was not one of Tunnicliffe's favourite subjects. He had measured drawings of a Turtle Dove and of a Collared Dove, as also of Wood Pigeons and a Stock Dove. I recollect exhibition paintings of both Collared Doves and Turtle Doves but one is left with the impression that he did not readily depart from his birds of sea and shore and marsh, and the hawks and falcons predatory on them.

Fancy Pigeons

This painting, which Tunnicliffe called, 'Angels and Archangels', is a particularly well-organised and satisfying design. A bed of columbines in the background is woven into a pattern of white, with the white heads and neck patch of the farther birds acting as a counterpoint to the main white areas in the plumage of the displaying pigeon. The fanned tail in the foreground is also nicely placed in relation to the ruffed necks behind.

BEE-EATERS AND MAGPIES

Although Tunnicliffe took less interest in little birds than in the larger ones, he was constantly sketching the small birds of garden and field. His collection of measured drawings included any small bird that fell into his hands. The Royal Society for the Protection of Birds used to ask him regularly to design a cover for their magazine when it was called 'Bird Notes', but for exhibition picture-making he rarely resorted to species smaller than those we illustrate here.

He did make a number of paintings of Magpies, two of which are included in this book. They are very common birds in Cheshire and in Anglesey. Their boldly patterned black and white plumage, with some of the black brilliantly glossed blue or green, and their strikingly long tapering tails make them attractive subjects for composition. Their chattering family parties in autumn are irresistible, although they are usually seen in ones and twos. Tunnicliffe, I think, never travelled outside the British Isles but he would turn his hand, when so commissioned, to producing pictures of foreign birds, for instance the set of tea-cards of African birds that he designed for the Brooke Bond tea company. In such circumstances he would, if he could, find a specimen of the bird in a private collection or a public zoo. Failing that, he would do the best he could by working with skins and photographs. I suppose the picture of Carmine Bee-eaters to have been inspired by seeing captive birds. He has certainly found it possible to capture the essence of these almost startling birds and has produced a fine picture of them. They are in a tight flock, very much as they are commonly to be seen in the wild where, like the European Bee-eater, they breed, feed and migrate in large noisy groups.

Carmine Bee-eaters

These Carmine Bee-eaters are of the race found in the southern part of Africa. They feed on insects and the northern species has the curious habit of riding on the backs of sheep or goats, or even of Kori Bustards, from which perch they make aerial forays to catch the insects disturbed as the animal (or bustard) moves forward. The southern variety of the bird, however, seems not to have developed this bizarre hunting method.

Spring Magpies

A pair of Magpies (left) on a misty day in very early spring, perched in an alder tree with one of the birds uttering its chattering call. There is a dense ground mist and the tops of the trees and bushes on the distant hillside emerge dimly to make a delicately patterned background. The newly-opened, male, stamen-bearing catkins on the alder introduce a vertical element in the design to balance the bold dark tail of the left-hand bird. The tiny crimson female catkins of the current year are perhaps not yet open but the alder cones, dry residues of last year's female catkins, their seeds shed, are to be seen in dark clusters.

In the Thorn Tree

Above we have another composition of a pair of Magpies, this time with a horizontal emphasis. The birds are perched in a hawthorn tree contorted by bitter wind from the distant grey sea in the background. It is autumn or, since the leaves of the hawthorn are already shed, perhaps early winter. It has been a good season for berries and the tree is crowded with the hawthorn fruits, or haws. They will very soon be consumed by birds of the thrush kind, the winter-visiting flocks of Fieldfares and Redwings. Magpies prefer an animal diet but they do eat fruit and berries as well.

BIRD PAINTING

The problems assailing artists are many and various. They depend to some extent on the nature of the chosen subject matter. At one extreme we may consider the painter of still-life. He can arrange his pots, pans, bowls, fruit and textiles in any manner that he may choose. He can fix the light direction and strength. Indeed his exercise in composition and design is completed before he starts to draw or paint. Above all he can be assured that the arrangement he has chosen will remain unchanged indefinitely. He can take his time in transferring his already completed aesthetic exercise to paper or canvas, the problems being reduced to those of technique in applying the right paint to the right place.

A portrait painter is in a similarly fortunate situation inasmuch as he can persuade, or hire, his subject to sit for long periods in one position in a pre-determined light. The next step, the transference to canvas, may be more difficult and call for more exactitude of draughtsmanship, and more human understanding, than in still-life painting but he has no need to work from memory or from a subject in motion, given a reasonably good model.

The landscape painter may well be troubled by unsettled weather, by changes in light, by movement of cloud or tide. He will often assert that a selected view will not look the same for two minutes together and indeed the good landscape painter may be regarded as painting light, weather and climate rather than topographical structure and detail. Nevertheless this structure and detail provide the foundation on which his observations are built and it, at least, will remain more or less constant.

At the other extreme from the still-life painter lies the painter of birds. The subject will rarely sit still. Unless it is a domestic or semi-domestic creature in a state of repose it will hardly remain in one position for more than a few seconds at a time. Moreover a wild bird is shy and not easily approachable. It can be seen well enough by means of binoculars or telescope but an artist has only two hands and it is impractical to operate binoculars, sketchbook and pen or pencil simultaneously. Time was when the artist went out with a gun and shot his bird and then made his picture at home from the corpse. This method had little to recommend it. At the best of times, it led to improbable and stilted attitudes; and now that man is becoming conscious that his overbearing fertility is tending to crowd other creatures off the earth, it is scarcely a socially or morally tenable course. Bird artists are too numerous and birds becoming too few.

The problem can be divided into two. The artist needs to know all there is to know about his subject so that when he draws a bird it looks like a bird and, moreover, like the species it is supposed to be. But it is not sufficient, except perhaps for zoological illustration purposes, just to represent a bird accurately. It is necessary that the artist study the behaviour of his subject, how it tends to place itself in relation to other birds and how it moves. Secondly, and of more importance, he must understand the nature of the birds setting and the inter-relationship between this and the bird itself.

To the end of his days Charles Tunnicliffe protested that he was as much likely to be excited by a shapely horse as by a shapely bird, but he had nevertheless devoted his middle and later years to becoming a painter of birds. He would scorn the notion that there is such a thing as a 'bird artist', protesting that he was an artist of general capabilities who chose, more often than not, birds as subject matter for his pictures; in spite of a lingering affection for the farm animals of his youth. In landscape he had less interest. A mown hayfield, to him, was not a subject for aesthetic or sentimental contemplation. He had too many memories from his youth of the farmer's anxious and practical consideration; when would the hay be dry enough to gather and carry home and would there be time enough to do it before it rained again?

In his book '*Bird Portraiture*' he remarks, 'one cannot draw a pose when it is not retained by the model for more than a few seconds.

'How then can we make drawings of the quick-moving bird? I know of only one way, and that is to watch, and watch, and watch again. Do not attempt to draw while you have the bird in view, but try to get accurate impressions photographed in your mind so that, when the bird finally disappears, you can get to work on your sketchbook and set down these impressions at once.' In a television interview in 1976 he said 'Get a sketchbook out and get as much of the live bird as you possibly can, and when he flies away from you, you go home and you have another sketchbook where you try to remember what you have seen.'

And indeed he practised as he preached. He was always drawing, either directly from life or from memory of life. In earlier days his sketchbooks were usually filled with drawings made in the field, though the sketches were in some cases elaborated subsequently, often with wax crayons. Latterly the boys of Manchester Grammar School, where he taught art during the war years, supplied him with specially bound books of water-colour paper. These books were kept in the studio and used, when he reached home, to make fine colour sketches from memory, aided by the outdoor drawings that he had just made in the rough book with pencil or pen and ink.

The strictly 'field' sketches were achieved with marvellous facility but with firm discipline. In making them he resolutely avoided using his memory of the bird from other occasions or his knowledge of what ought to be there if only he could see it. He confined himself rigidly to what he could actually see. In consequence many of these sketches, particularly of wild birds, are very slight but nevertheless brilliant notes that convey the very life of the bird. Note, for example, the drawing (page 120) of a Heron in repose. We may suppose that it moved, or flew away, before the drawing could be elaborated. But in this sketch of a moment he has recorded the shape, the posture and, above all, the disposition of the tonal areas and, unmistakably, we have the essence of Heron in a few lines. A bird in constant motion would be more difficult and a page might be devoted to a variety of poses,

Field drawing of resting heron

Eider Ducks

Red-legged Partridges

some merely indicated, others more substantially drawn, the whole becoming a collective summary of what the bird-watcher is apt to call the 'jizz' of the bird. The sketches of Eiders and of Red-legged Partridges (opposite) are of this kind. Tunnicliffe was very adept at recording, in this way, the various changes in attitude of a bird in even rapid movement. His slightest sketches could be sufficient to enable him to make convincing pictures of the fleeting moment.

When he was able to draw at greater leisure, on those happy occasions when a wild bird posed within range for a sufficient time, or when he had access to captive birds, then the 'field' sketch became more elaborate and detailed. The two drawings of owls that we reproduce, of a Long-eared Owl and of an Eagle Owl (below) were both made from captive birds, the former at Penrhos reserve on Anglesey and the latter probably at Chester Zoo. Birds of prey were frequently accessible under favourable conditions and in the drawings of a Gyrfalcon (page 123) we see, besides summary notes of postures, some more detailed portrait studies of the bird, of heads and bills and feather detail. Again, when a bird could

Pencil sketch of Long-eared Owl

Study of an Eagle Owl in captivity

be studied at leisure the drawing might be supplemented by written notes of colour and structure. The drawings of Lady Amherst's Pheasant and of a Black Swan (opposite) display this feature of his field work.

The 'studio' sketch that frequently followed this field work would be based on the field sketch and supplemented by memory and imagination. Whereas the field sketch was nearly always no more than a pencil drawing, his work in the studio would normally be in colour and, if the occasion suggested it, they would often record not simply the bird but arrangements and juxtapositions that could be the beginnings of an idea for a picture.

For more detail of eye and bill, foot and feather Tunnicliffe made carefully measured drawings in colour, similar to that of the Black-tailed Godwit (page 136) but life size, of any dead bird that came into his hands; showing the bird seen from above, from below, wings spread, wings closed, and with special drawings of heads and feet. He was not averse to the study of museum skins (though apt to refer to them disparagingly as 'useless sausage-shaped things') but he found much more information in his own drawings. By the end of his life he had a collection of more than 300 of these. They acted as a supplement to his sketches and ensured accuracy of detail. He remarks in '*Bird Portraiture*' that you have to bear in mind that 'there is always a fierce ornithologist just around the corner, ready to pounce and rend you' if you get something wrong. Museum skins have their limitations, black and white photographs confuse pale and dark with light and shade, and colour photographs often report colours falsely. Tunnicliffe's measured drawings, made with the care that he devoted to them, were unquestionably accurate in every respect.

The sketches and the measured drawings occupied a great deal of Tunnicliffe's time. Together they enabled him to draw and paint any bird he had studied in exact shape, posture and colour, and in any necessary detail. A bird is not a picture, however, though there may be many who think it is. It is a creature of intrinsic beauty, but an exact representation of a living being on which nature has conferred beauty does not of itself constitute a work of art. Tunnicliffe to quote him again, took the view that a work of art should have its 'own particular claim to be beautiful, not because it slavishly imitated the form and colour of a bird, but because it has used the bird and controlled it to make a new beauty'. From sketch, measured drawing and memory how then did he proceed to the production of the finished painting?

In some instances the bird, or group of birds, in the posture and setting he had actually seen, would suggest an idea for a picture. The group might be so near to what he considered would make a good painting that the studio sketch might already be very close to what eventually emerged as the finished product. In other instances modification might be needed. I remember, on looking through his sketchbook on one occasion, remarking of one particular page, 'That would make a good picture, wouldn't it?' 'Yes, and it will, if I can get a good arrangement.' The arrangement in that case was present in the sketch but was not quite right.

Pencil studies with colour notes of captive Lady Amherst's pheasant

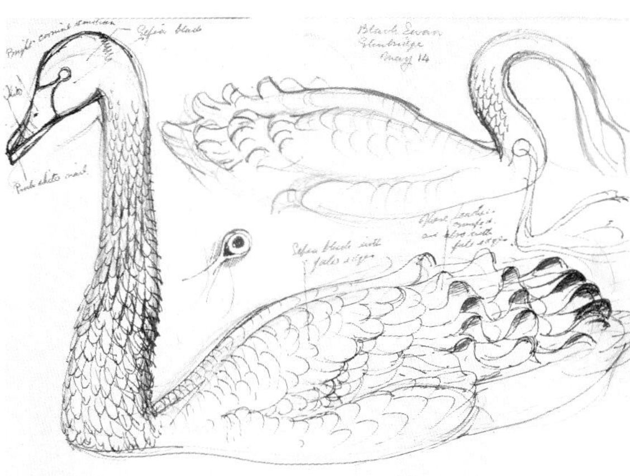

Black Swans, sketched during a visit to the Wildfowl Trust at Slimbridge

Above and right: Pencil drawings of a Gyrfalcon, with anatomical detail, probably made from an injured bird at Penrhos Nature Reserve near Holyhead in Anglesey

There were also cases where the idea for the picture did not emerge directly from something seen but was the product of imagination; the response to a self-questioning, 'suppose this or that particular bird were to be seen in such and such a situation?' Some of Tunnicliffe's wall pictures were painted to commission and, while the patron might sometimes want a picture more or less like one he had seen – already sold – in a gallery, it was more usual that he would ask for a specific bird and leave it to the artist to devise a setting and a suitable pictorial composition. In some of these private commissions he was able to design a picture that was not only a delight in itself but, in collaboration with the patron, harmoniously related in size and colour pattern to a particular position in a particular room, contributing to the pleasing ambience of the whole room as a place to live in.

Whatever the circumstances, the first step that Tunnicliffe took was to surround himself with all relevant reference material in the shape of sketches and measured drawings of the bird in question. To this would be added sketches of other items to be included in the picture, as an appropriate setting. He then produced what he called a 'composition sketch'. This would be a quite small, very rough painting in water-colour, usually in gouache, in which the bird, or birds, and the surroundings were boldly indicated in relation to the picture rectangle. There was little or no attention to drawing or to natural history, the sketch was simply a statement of a pattern in line and in areas of tone and colour. A few of these sketches from which

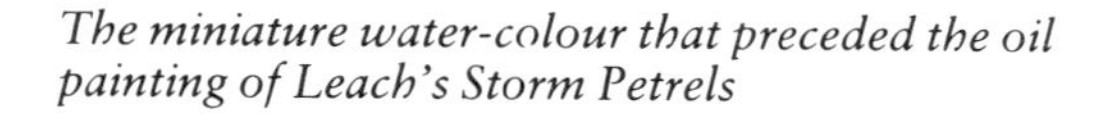

The miniature water-colour that preceded the oil painting of Leach's Storm Petrels

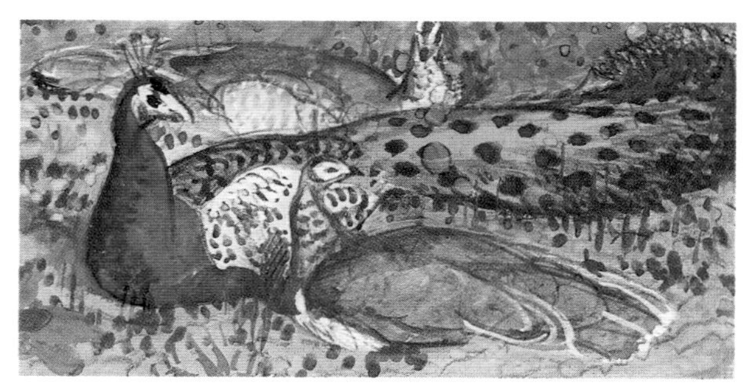

Composition sketch in anticipation of the picture in oil of resting peafowl

pictures earlier in the book were directly derived are reproduced here and it will be appreciated that the merits of the picture, as a picture, are already apparent in the composition sketch, the detail of exact representation being almost subordinate to the aesthetic pleasure to be derived from contemplation of the finished picture.

Of course minor, sometimes major, adjustments were made at later stages, but the essence of the picture, as a picture, lies in the composition sketch. Consider the sketch (above) of the Leach's Storm Petrel painting (frontispiece) and note how differently the birds or, for that matter, the waves, are arranged. Nevertheless here in the sketch are the bones of the picture as it was to become. Sometimes the changes made at later stages are of interesting significance. The sketch (opposite) of the resting group of Peafowl that preceded the oil painting (page 106) shows the necks of the three birds as verticals. In the final version the neck of

Design for 'Angels and Archangels'

the far Peahen has been curved to harmonise with the curve of the tail of the Peacock, a designer's refinement that is subtly effective. The arrangement of the pigeons in the sketch (above) foreshadows directly that of the painting he called 'Angels and Archangels' (page 113) but the background of white columbines has been added as a delightful afterthought.

Sometimes the composition sketch is an almost exact anticipation of the end result as we can see if we compare the sketch (page 126) with the picture of Mute Swans (page 54). There is little difference between the sketch (page 126) and the final picture of the Blue-Eared Pheasants (page 100).

Many of the modifications may have been effected at a later stage in the preparation of the picture, but in some instances Tunnicliffe produced several composition sketches before he was satisfied. Thus the composition sketch (page 126) for the picture of Barnacle Geese (page 43) is one of at least three on the same theme in which the essentials of the picture are unchanged but the arrangement of water, sand and grass and the postures of some of the geese are varied. None of these three composition sketches is quite identical in lay-out with the finished picture. On the other hand, the composition sketch (page 126) that preceded his painting of Night Herons (page 70) required much less adjustment as did the one of the Peregrine (page 126) that foreshadowed the finished picture (page 88).

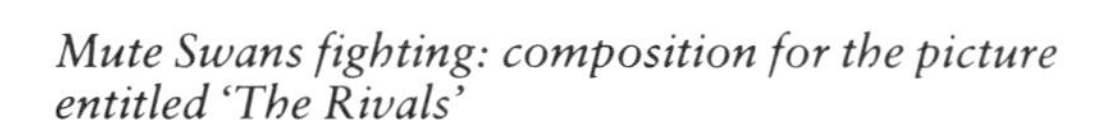

Mute Swans fighting: composition for the picture entitled 'The Rivals'

One of several layout ideas considered for the painting of Barnacle Geese

Blue-Eared Pheasants, the design for one of Tunnicliffe's most successful Japanese-style paintings

Night Herons

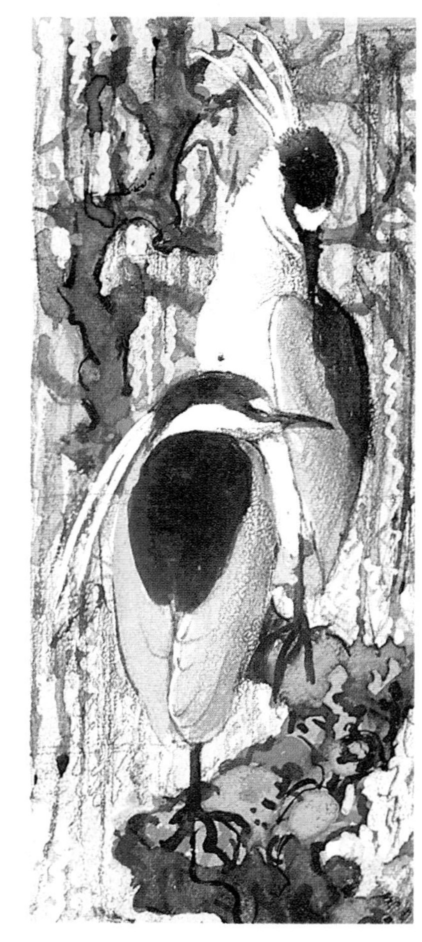

Composition sketch for the Royal Academy Summer Exhibition submission 'Tiercel Perch'

Tunnicliffe himself thought highly of these composition sketches. They were pasted onto backing sheets, preserved in portfolios, and were no doubt of great help in the discussion of possibilities with a patron who had no very clear idea of what he wanted. A study of the portfolios is absorbing indeed, for it is in them that the man's unique genius is to be found. Here is the difference between this artist's work and that of the zealous recorder merely of ornithological detail.

Once more or less pleased with the composition as set out in the sketch, Tunnicliffe would enlarge it to the size of the finished picture, usually on tracing paper. The sketch is sometimes squared up to facilitate this operation. The monochrome work on tracing paper was then developed into a finished drawing, with zoological detail stated and checked from appropriate sketches and measured drawings, and the niceties of linear design carefully worked out. When this had been done and he was entirely satisfied, the drawing would be traced onto the final sheet of water-colour paper carefully chosen in tone, hue and surface for the particular task. Tunnicliffe set great store by the purity and simplicity of water-colour washes, maintaining, rightly, that the more directly water-colour is applied the better it will look. He was thus reluctant to make alterations to his drawing on the final paper because pencil and india rubber tend to damage the surface of the most robust of water-colour papers. When this happens, washes of colour will not lie so beautifully, however carefully applied. In other words transparent water-colour has its own intrinsic beauty which should be assiduously cultivated in order to produce the finest painting.

Thus the drawing, quite finished in natural history, perspective and design, would be lightly traced onto the selected paper, which was wetted, stretched onto board and allowed to dry. The painting was then executed rapidly in full rich washes – if possible attaining the desired tonal value in a single wash – using wet or dry areas of paper as might be appropriate. As far as possible he used transparent water-colour but he was not averse to finishing touches of gouache. Some of his highly decorative patterns of light on dark were attained by careful use of masking fluid. Occasionally a texture of light on dark would be achieved by scraping with a sharp knife on a hard rough paper. But as far as possible the finished painting was carried out by rapid, direct painting in transparent water-colour.

I have already suggested that what distinguishes a Tunnicliffe painting as a work of art from a mere study of a bird, however exact a replica, lies in the composition, in the design of the painting within the picture rectangle. This question of design is to some extent a matter of feeling, of knowing what will look right. It is also a matter of taking into account certain rules and well established principles with which he, as a successful student of the Royal College of Art, was thoroughly familiar; of balance between straight line and curve, light and dark tones, cold and warm colour. Also, especially in Tunnicliffe's designs, a system of point and counterpoint, a concentrated major interest in the picture echoed in a diffuse and minor key elsewhere within the picture area. Once equipped with a series of sketches and a measured drawing of the appropriate corpse – and an ability to draw well – the production of a good representation of a bird was a matter almost of routine. The difficulty, and the genius, lay in recognising in nature, or devising, the 'good arrangement'.

Let us now examine the antecedents of some of his paintings in the sketchbooks, in the measured drawings, and in further examples of the composition sketches in which the fundamentals of the design are established.

Frequently a painting appears to spring directly from the recognition that a certain group of birds, in the setting as observed would, with little or no modification, make a good picture. For example, consider the development of the picture of the two resting Shelduck (opposite). This can be followed in stages, from a pencil drawing in the field (below) of the two individual poses (only one of which shows the reflections that lie at the root of the final design) through a studio sketch (below) which is clearly already an incipient picture in which the balance

Field study in pencil

Studio coloured sketch

Composition sketch for 'Shelduck at Rest'

Shelduck at Rest

A pair of Shelduck rests in still shallow water. One stands on the bottom, the other seems to float. Both have their bills tucked into their back feathers. The reflections are a perfect mirror-image of the birds, and together give rise to abstract shapes of striking beauty. One of the birds has its eyelids open but the eye is closed by the white nictitating membrane. The curves of the sandspit complete by artifice a pattern otherwise designed by nature.

of shapes, colours and tones makes a delightful abstract pattern, to a composition sketch (page 128) in which the relative position of the birds has been altered only slightly and there is a suggestion of a possible pattern in the background. In the final painting there is a further minor adjustment to the relative position of the ducks and the simple pattern of the background has been crystallised.

There are a number of other instances where this almost direct transference of something seen to the final picture may be detected among the paintings reproduced in this book. The painting of Mute Swans (page 49) for example, that Tunnicliffe dubbed 'Preening in the Sedges', is clearly developed from a sketch (opposite) he made at Cemlyn Little Pool. His 'Green, Gold and Dun' of Lapwings, Golden Plover and Dunlin (page 62) represents an almost direct transference from a studio sketch (below) of a similar group seen beside the Cob Pool. Charles's notes on the sketch are interesting:

On a spit of land projecting into the lake. Lapwings, Golden Plover and Dunlin crowded onto the tongue of land. Dunlin were still all in winter plumage and all asleep. Lapwings were showing signs of change to breeding plumage. Goldens still in apparent winter plumage except one which had some black on breast. The air was still and the reflections were not disturbed. The water was reflecting the brown sides of the Cob and the birds were against this brown background.

Lapwings, Golden Plover and Dunlin

It is also interesting that in the first version of this picture, which was exhibited at the Tryon Gallery in London, the Dunlin were 'all asleep', as he notes and recorded in the sketch. The version we reproduce was painted subsequently as a private commission and one of the few differences between this and the original is that one or two of the Dunlin have raised heads.

The picture of Black-headed Gulls (page 81) that Tunnicliffe called 'The Ninth Wave' is also directly developed from a very rough sketch (opposite) made at Porth Nobla, but in this case extensive reorganisation is evident in the composition sketch (opposite) to make a more suitable balance between dark

Mute Swan and cygnets at Cemlyn little pool

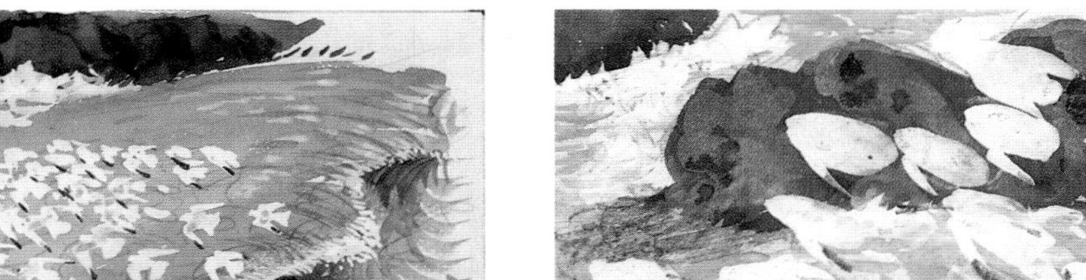

Studio sketch of gulls at Porth Nobla

Composition sketch for 'The Ninth Wave'

rock and light wave. Again it may be interesting to record Tunnicliffe's notes, written on the original sketch:

> *Flocks of gulls, mostly Blackheads with a few Common rising to clear a big wave. They let the wind lift them clear and settled immediately the wave had passed beneath them.*

The Common Gulls, just recognisable in the sketch, have been omitted from the finished painting.

The picture of two resting Herons (page 71) has a directly antecedent sketch made at Cemlyn (below). The posture of the sleeping birds has hardly changed but Tunnicliffe noted 'Two Herons sheltering behind a pile of bricks and the high wall bounding Cemlyn House'. In the picture, however, the pile of bricks and the wall

Studio sketch of sleeping herons

Composition sketch for 'Heron Cove'

Studies of first winter Sandwich Tern
made at 'Shorelands'

Studio sketch of juvenile Sandwich Terns

have been replaced by an elegant setting of rocks and seaweed and this change had already been adumbrated in the composition sketch (page 132). 'Heron Cove' evidently bears little relation to what the artist had actually seen before he conceived the picture.

The group of Sandwich Terns (page 80) is almost certainly developed from a sketch (page 133) of a group of juveniles. These are drawn resting on dark rocks and include among them a Black-headed Gull. In the finished picture the terns are resting on a pattern of mud pimpled by lugworm casts and the gull has been omitted. There can be little doubt that the other series of sketches (page 133) of a juvenile Sandwich Tern that was found injured and taken to Shorelands provided some of the essential knowledge that went into the preparation of this painting.

The picture of Pheasants (opposite) in which a Cock Pheasant displays to a hen appears to be a picture imaginatively composed. But the sketch (below) shows a study from life of a displaying cock with the note:

> *Cock displaying in front of hen bird drops flank feathers and lifts the open tail sideways so that the upper surface is towards the hen. Direction of the tail is at an angle to the direction of the body.*

Only by using the knowledge contained in these sketches and notes would it be

Colour sketch of Cock Pheasant displaying

Post mortem details of Cock Pheasant

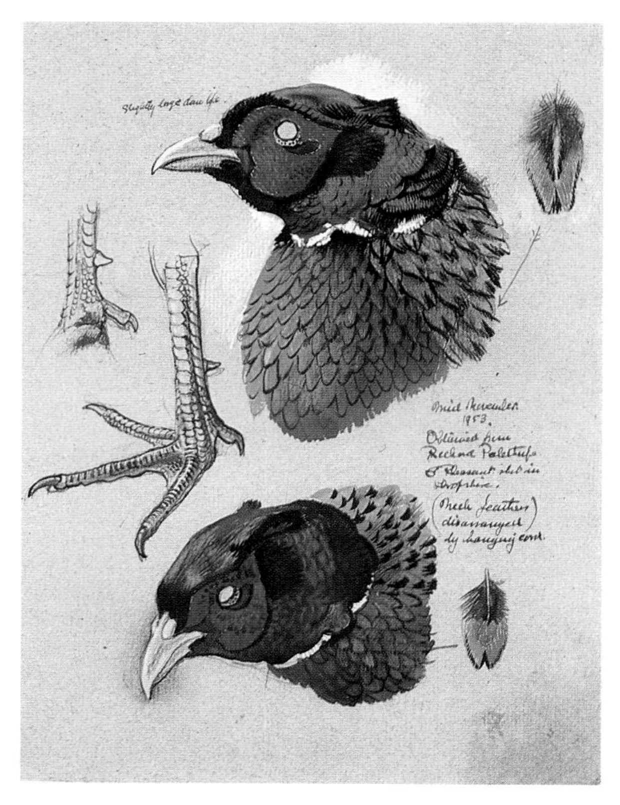

Pheasant Display

A cock Pheasant displays to a hen. The posture is very like that of the domestic cockerel, with tail fanned and nearside wing spread and drooped, the whole body tilted towards the female so that she misses none of the glory. It is early spring, as manifest also by the surrounding vegetation in which appear lesser celandines and primroses among last year's dead bracken and fallen oak leaves. The female is, by comparison, soberly clad against the splendour of her mate but when she comes to hatch a clutch of eggs the value of her mottled brown colouration, as a concealment, will become apparent. She is barely visible as she sits motionless in her rudimentary nest among dead leaves on the ground.

Scaled down measured drawings of Black-tailed Godwit, from one of the sketch books

Black-tailed Godwit wading

possible to make a convincing picture of such a display; no matter how much knowledge one might have of the bird in a more normal posture, and in whatever detail. The other sketch (page 134) shows drawings of a Cock Pheasant's head made from a bird shot in Shropshire.

Midsummer Godwits, as Tunnicliffe named the fine picture (page 138) of a party of Black-tailed Godwits coming in to land on a pattern of pools and mud, appears to be an entirely imaginative composition, no doubt based on something often seen and carried in memory. There are drawings of a Black-tailed Godwit, made from a corpse, in his sketchbooks, not life size as he usually made these drawings but to scale and with actual measurements recorded (opposite). These show a Godwit still in partly red plumage with a pale face as in the finished picture. Doubtless they also provided the ornithological detail for the flying birds. Tunnicliffe had many splendid sketches of these birds, made as they visited the Cob Pool in spring and summer passage. One of these shows a bird wading (opposite) and the other comprises a number of posture studies (below) in some of which the birds are shown with spread wings.

The painting of a Little Owl (page 140) is, so far as the bird is concerned, simply

Posture studies of Black-tailed Godwits

Midsummer Godwits

Six Black-tailed Godwits are in the process of alighting on an area of blue shallow water and purple spits of mud. In flight the distinction between Bar-tailed and Black-tailed Godwits is at once apparent, obscure though it may be in the standing bird. The bold pattern of black and white in wings and tail is totally lacking in the Bar-tailed species which in flight recalls the Curlew's subdued design. 'Midsummer Godwits' was Tunnicliffe's own title for this picture. These birds appear briefly at Malltraeth in April but soon disappear to their breeding areas. They appear again towards the end of July, one of the earliest of the waders to start the move towards winter quarters. It may be noted that all these summer birds have pale patches on the face, possibly indicative of the beginning of the moult to winter plumage.

a portrait of a Little Owl. Tunnicliffe had a great many drawings of this species, having reared at least one that was brought to Shorelands. Of the sketches in colour (below) he notes:

> *Juvenile Little Owl. Brought in a paper bag by two boys from Capel Mawr. Could not fly when it came to us. Fed on rabbits and dead birds.*

The pencil drawings we have reproduced (page 141) are also made from life, possibly of a wild bird seen resting in daylight. These are probably more directly related to the figure in the painting whose merit derives from the association of the

Studies of fledgling Little Owl, made at 'Shorelands'

bird with the ancient tree, and the subtle relationship of line, tone and colour between the bird and its environment.

It is illuminating to examine the composition sketches in two instances where the picture has a low horizon. The setting is not the usual one of a few square yards of ground on which the observer looks down from a relatively short distance, but is a distant landscape of land and water. Consider the sketch

Little Owl

This painting of a Little Owl is almost as much a picture of an ancient, contorted and dead tree, on which the owl is perched and to which as much attention has been given as to the bird. The Little Owl, a species introduced into the British Isles last century, is given to sitting perched in the open in the dusk or even in broad daylight. The subtleties of design merit study: note how the branches and dead leaves in the hollow provide counterpoint to the owl's feather pattern.

that formed the basis of the painting of a Snowy Owl (page 90). It will be recalled that this painting was stimulated by Tunnicliffe's being brought the body of a female Snowy Owl that had been found at Mynachdy near Carmel Head in the north-west corner of Anglesey. He made a fine measured drawing of the corpse, which, he noted, had been shot and one leg had been removed from it. The painting, we must suppose, represents what might have been seen before the shot was fired, the owl haunting that wild and rugged headland with the West Mouse and its lighthouse and the Skerries out to sea. Tunnicliffe, in the nature of things, could have had no sketches of that particular bird but he did have a number of sketches of Snowy Owls made at Chester Zoo (see, for example, '*A Sketchbook of Birds*', pages 92-93), and we reproduce a pencil sketch (below) made very probably on the same occasion at Chester. The composition sketch shows the owl in a very similar posture to that in the pencil sketch but the bird is turned partly away from the observer. In the finished picture, probably completed after a visit to

Field sketches in pencil of a Little Owl

Pencil sketch of captive Snowy Owl

Rough design for picture of Snowy Owl

the spot, there is a more precise delineation of the rock forms and both the owl and the landscape are illuminated by clear sunshine so that the bird is not so much a white blob against a dark ground. The composition, in other words, has been balanced and improved at the cartoon stage.

We may also consider the composition sketch (below) that foreshadowed the painting (page 52) of Mute Swans flying over the Cob at Malltraeth, a scene often contemplated from Tunnicliffe's window at Shorelands. This sketch might almost be considered better than the finished picture, which not infrequently happens when an artist proceeds to elaborate a fortunate sketch. There are seven swans in both cases but they have been somewhat re-arranged. In the sketch the white upper surfaces of the wings of the three leading swans form the focus of attention.

Composition sketch for 'Mute Swan Flight'

In the finished picture the illuminated wings have been dispersed more widely and the focus is less emphatic. But perhaps more interestingly, the finished picture has, as of course is usual, much more detail not only in the swans but in the landscape. As a result the picture has suffered in the loss of the bold silhouette of blue hills and the contrast between the dark landscape and the pale pool. The pool is almost lost in the final picture, the painting of detail in the hills, the Bont Farm lands and the foreground Cob having destroyed to some extent the original pattern. The attractions of a recognisable landscape has been achieved by the loss of a more incisive and striking design.

It may be appropriate to conclude with an assertion of the writer's opinion that Charles Tunnicliffe was one of the greatest, perhaps the greatest, painters of pictures of birds that has ever lived. This opinion is based not so much on the belief that he could represent the living bird more accurately and convincingly than anyone else – though he could – but on the conviction that he combined the knowledge and enthusiasm of a naturalist with the trained skill of an accomplished draughtsman and painter *and* with, and this is the point, a feeling for composition, design, decoration, pattern, call it what you will, that in his best pictures puts him quite outside the run of 'bird artists' and into an honourable place among artists at large.

It may well be that his standing as an artist has not been assessed as fairly as it should have been simply because he chose birds so frequently as his subjects. His work is understood, admired and respected by a considerable following, found perhaps mainly among those who love birds rather than among those who set fashions of contemporary art appreciation. Paintings of birds are apt to be dismissed as mere zoology by those who have no special regard for such subject matter, and it is possible that a proper appreciation of bird paintings, as great art, can never be fully realised in an atmosphere of experiment, abstraction and subjective emotional involvement.

FURTHER READING

Portrait of a Country Artist by Ian Niall
(Victor Gollancz 1980)
Had Tunnicliffe done nothing other than illustrate books, he would have rated high and his name would be remembered. But his achievement as a wildlife artist, and in particular as a bird artist, is of major importance. Turning to the study of birds in the 1930s, he devoted months at a time, year in year out, to observing and sketching his favourite species, first on the Cheshire meres and nearby moors, then in Anglesey, his home from 1947 until his death in 1979.

This book attempts to illustrate a sample from Tunnicliffe's enormous output of work in its various facets. The superb early etchings are well represented, as are his fine wood-engravings from the late 1930s. His watercolours, which became a regular feature of the Royal Academy Summer Exhibitions, are reproduced in colour, together with some of his sketches and measured drawings of birds. Particularly beautiful are the marvellous oils on cloth from 1938, little known now but surely some of his most lovely pictures.

Ian Niall's affectionate portrait of his friend describes the life of a warm-hearted, down-to-earth countryman, a totally dedicated artist whose genius is evident in his depiction of scenes from country life and in the most exquisite representation of wild creatures.

A Sketchbook of Birds Introduced by Ian Niall
(Victor Gollancz 1979)
In his introduction Ian Niall describes how his research work for *Portrait of a Country Artist* led to the prior publication of a book of Tunnicliffe's sketches:

As the pages of my notes increased, I felt that it was time to look for a publisher. My friend and fishing companion, David Burnett of Victor Gollancz, needed no second invitation to come up from London and talk about the book with Charles and myself. When he arrived at my house I proceeded to talk him into the ground about The Life and Art of Charles Tunnicliffe. *On the journey over to Anglesey he kept falling asleep as I continued at length to describe the marvels that Tunnicliffe had shown me, including his drawings for the original wood-engravings for* Tarka, *and his extraordinary 'bird maps', large scale detailed studies of hundreds of birds, some of which I thought might go into my book. On arriving at 'Shorelands', the house at Malltraeth where Tunnicliffe lived for thirty years, we looked together at the 'bird maps' and then I once again took up with Charles the thread of our previous discussion. My publishing friend had fallen into a kind of reverie over some sketchbooks he had found casually stowed in a cupboard. It seemed to me that he was pondering technical problems. I was afraid that he would say that my book could not be illustrated in the generous way I had hoped.*

Suddenly he remarked that here was treasure trove. Tunnicliffe's bird sketches were something the world must see.

A Sketchbook of Birds was followed by *Sketches of Bird Life* and later by *Tunnicliffe's Birds*, a magnificent book of the measured drawings with a commentary by Noel Cusa, all published by Victor Gollancz Ltd.

Tunnicliffe's Countryside by Ian Niall
(Clive Holloway Books 1983)
Charles Tunnicliffe's name was brought to the world's attention in 1932 with the appearance of the first illustrated edition of Henry Williamson's *Tarka the Otter*. Publishers of country books were quick to respond to his talent and encouraged him to a prodigious output that was to run side-by-side with his brilliant work as a bird-portraitist.

In 1942 The Studio published the artist's own semi-biographical work *My Country Book* which was followed by *Mereside Chronicle* and the final masterpiece, *Shorelands Summer Diary*, in which he describes his life on Anglesey. *Tunnicliffe's Countryside* is based on these three books and upon a selection of his illustrative work; chosen to reflect the genius that he brought to the task, by seeing into the minds of the many country writers whose world he knew, and understood, so well.

After *Tarka the Otter* came Williamson's *The Lone Swallows, The Old Stag, The Peregrine's Saga* and *Salar the Salmon*. He was commissioned for H.E. Bates, and Alison Uttley would have no other illustrator for her accounts of rural childhood.

In 1965 Country Life Books brought out Ian Niall's *The Way of a Countryman* with Tunnicliffe illustrations and this was the first of four books on which the two collaborated; an association that provided the inspiration for *Portrait of a Country Artist*, Ian Niall's account of the life of his friend. *Tunnicliffe's Countryside* is a further tribute to his perceptive observation, and unfailingly accurate portrayal, not only of the birds, but of all that lived about him.

Mereside Chronicle by C.F. Tunnicliffe
(Country Life 1948)
Apart from cherished memories of childhood, preserved in his sketches of the village and farm at Sutton Lane Ends in Cheshire, Charles Tunnicliffe confirmed a distinct environmental inspiration in the later stages of his life. He could not return to the farm after his studies at the Royal College because it had been sold, so he had gone to live in nearby Macclesfield. He had married his first love, who supported him in everything he did and worked as a teacher of art and craft herself; while he threw himself into frenzied activity as a freelance commercial artist. At the same time he was doing book illustration and taking the first steps

towards bird portraiture. It was as a by-product of his field work, sketching birds, that *Mereside Chronicle* came about. Before this Tunnicliffe had gone to the canal banks, to sketch and draw subjects that were in the main scenic and as near to landscape as he would ever get. But the Cheshire meres were something different. They were not just escape from the unexciting background of the housing estate and the treeless avenue in which he lived. They were the haunt of waterfowl, the natural scene in which the grebe or swan swam in open water and nested in the reeds. The young artist had acquired a motorbike and took himself off, whenever he got the opportunity and deadlines with his advertising agency would permit, to sketch at Bosley Reservoir, Gawsworth, Radnor, Siddington and Capesthorn. His first concern was to collect sketches of the various species for reference; until one day he realized he had a full record of the bird life of the meres and pools; with all the seasonal changes and their effect on the birds and their surroundings. Encouraged by *Country Life*, he increased the number of his visits and completed a kind of diary of a twelve month, *Mereside Chronicle*, which delighted his fans when it was published in 1948. Quite plainly a part of it, dealing with Tunnicliffe's visits to the highlands of Scotland, was a kind of make-weight; but nevertheless it in no way detracted from the quality of the whole. The book contained no less than 207 illustrations, plus pages of maps, and a dust jacket which was, like the rest of the work, in monochrome. The world of the mereside was part of Charles Tunnicliffe's beloved native country, a scene to which he was irrestibly drawn while he lived in what was, for him, a dreary suburban atmosphere.

Shorelands Summer Diary by C.F. Tunnicliffe
(Collins 1952. Reprinted by Clive Holloway Books 1984)
One of the advantages of being a writer or an artist is that the person who has opted for either of these professions may choose where he settles. For the writer it is sometimes to his advantage to be distanced from his subject and benefit from significant recall. The artist will refresh his impressions and sometimes benefit from having his subjects on his very doorstep. There was certainly a feeling that to live close to his subjects was essential when Tunnicliffe took his wife on a holiday to the island of Anglesey; for it was at around this time that some of his field work on the Cheshire meres was being culled for *Mereside Chronicle*. The happy couple fell in love with the place. It had everything it seemed; a farmland landscape, distant mountains, seacliffs, strands on the tidelines where waders trilled and cleeping oyster-catchers worked over wet seaweed. More than this, on the mouth of the Cefni estuary, at a sleepy Welsh village called Malltraeth, there was a house, tucked away, but not a stone's throw from the shore. It seemed 'the stuff that dreams are made on' and Charles Tunnicliffe, after he had reconnoitred the place from a respectable distance, hastened to make the acquaintance of the proprietor of the nearby Joiners Arms. He persuaded the landlord to get in touch with him if and when that quiet house on the shore came up for sale.

Soon after their holiday the Tunnicliffe's received the news they had hoped for, and lost no time in buying their dream. *Shorelands Summer Diary*, a masterpiece, was the result of that decision. It is not only an account of the move and the conversion of a room into a studio but of the dream coming true as Tunnicliffe sat behind his window and watched the seabirds come in and go back with the tide. Here were all the waterfowl he could ever have wished for, and more besides, shags fishing from the rocky outcrops, peregrines at South Stack, the hawks and owls he drew so well; rarities that kept him out field sketching in sun and shower, summer and winter. Often he was out sketching for as much as ten hours at a time; after which he would drive himself to catch up with other important items, commissions, and deadlines for advertisers. Rarely did the Tunnicliffes go away and leave this wonderful place. When they did it was almost invariably to gather further reference, to get closer to the elusive bittern or the dainty, shy avocet. More than half Tunnicliffe's working life was to be spent here. Both he and his wife ended their days at 'Shorelands' and *Shorelands Summer Diary*, an account of a single season, could be called the summer time of their lives. It was beautifully produced by Collins with sixteen colour plates and 185 vignetted scraper-boards; and it has become a much-prized collector's piece for those who appreciate this particular fragment of Tunnicliffe's life's work.

PICTURE INDEX

Bee-eater, Carmine *Merops nubicus* 115
Buzzard, *Buteo buteo* 85, 89

Coot, *Fulica atra* 21, 47
Curlew, *Numenius arquata* 59, 64

Dunlin, *Calidris alpina* 62, 69, 79, 130
Duck, Tufted *Aythya fuligula* 21, 30

Eider, *Somateria mollissima* 32, 120

Fowl, Domestic *Gallus gallus* 95, 97, 98, 104, 108, 109

Godwit, Bar-tailed *Limosa lapponica* 69
Godwit, Black-tailed *Limosa limosa* 68, 136, 137, 138
Goldeneye, *Bucephala clangula* 31
Goose, Barnacle *Branta leucopsis* 43, 126
Goose, Canada *Branta canadensis* 46, 47
Goose, Chinese *Anser cygnoides* 39
Goose, Red-breasted *Branta ruficollis* 42
Goose, Whitefronted *Anser albifrons* 37, 40, 44, 45
Goshawk, *Accipiter gentilis* 92
Grebe, Great Crested *Podiceps cristatus* 21
Grouse, Black *Tetrao tetrix* 105
Gyrfalcon, *Falco rusticolus* 91, 123
Gull, Black-headed *Larus ridibundus* 78, 81, 131
Gull, Common *Larus canus* 76
Gull, Great Black-backed *Larus marinus* 76

Heron, Grey *Ardea cinera* 71, 120, 132
Heron, Night *Nycticorax nycticorax* 70, 126

Lanner, *Falco biarmicus* 93
Lapwing, *Vanellus vanellus* 62, 66, 67, 130

Magpie, *Pica pica* 116, 117
Mallard, *Anas brachyrhynchos* 22
Mandarin, *Aix galericulata* 19
Moorhen, *Gallinula chloropus* 23

Owl, Eagle *Bubo bubo* 121
Owl, Little *Athene noctua* 139, 140, 141
Owl, Long-eared *Asio otus* 121
Owl, Short-eared *Asio flammeus* 87
Owl, Snowy *Nyctea scandiaca* 90 ,141
Oystercatcher, *Heamatopus ostralegus* 79

Partridge, *Perdix perdix* 102
Partridge, Red-legged *Alectoris rufa* 103, 120
Peafowl, *Pavo cristatus* 106, 107, 124
Peregrine, *Falco peregrinus* 88, 126
Petrel, Leach's Storm, *Oceanodroma leucorrhoa* Frontispiece, 124
Pheasant, *Phasianus colchicus* 99, 134 135
Pheasant, Blue-Eared *Crossoptilon auritus* 100, 126
Pheasant, Elliot's *Syrmaticus ellioti* 111
Pheasant, Lady Amherst's *Chrysolophus amherstiae* 101, 123
Pheasant, Reeve's *Syrmaticus reevesii* 110
Plover, Golden *Pluvialis apricaria* 62, 130
Plover, Ringed *Charadrius hiaticula* 77, 79
Pigeon, *Columba livia* 113, 125
Pintail, *Anas acuta* 28, 29, 33
Poultry, *Gallus gallus* 95, 97, 98, 104, 108, 109
Puffin, *Fratercula arctica* 83

Redshank, *Tringa totanus* 65, 79
Ruff, *Philomachus pugnax* 63

Shelduck, *Tadorna tadorna* 24, 25, 41, 128, 129
Shoveler, *Anas clypeata* 34
Snipe, *Gallinago gallinago* 61
Swan, Bewick's *Cygnus columbianus* 53
Swan, Black *Cygnus atratus* 57, 123
Swan, Mute *Cygnus olor* 49, 51, 52, 54, 56, 126, 131, 142
Swan, Whooper *Cygnus cygnus* 55

Teal, *Anas crecca* 27
Tern, Common *Sterna hirundo* 73, 75
Tern, Sandwich *Sterna sandvicensis* 80, 133
Turkey, *Meleagris gallopovo* 104, 108
Turnstone, *Arenaria interpres* 59

Wigeon, *Anas penelope* 26, 35

ACKNOWLEDGEMENTS

The publisher acknowledges with gratitude the assistance and co-operation of the Estate of the late C.F. Tunnicliffe and the various members of his family whose hospitality and encouragement have been such an important factor in the conception and production of this book.

A number of the pictures reproduced are owned by people living on the island of Anglesey. John Smith, an authority on Tunnicliffe's work resident on the island, has been responsible for negotiating permission to photograph these paintings for our use and has, singlehandedly, overcome the problems of organising sessions whereby several paintings could be photographed at the same time. This help has been of considerable value and we are most grateful to him.

Ian Niall, Tunnicliffe's biographer and a good friend, has given invaluable advice and practical help with this project, not only in the months leading up to publication but since it was first mooted some three years earlier.

Most of the pictures reproduced here are owned by private individuals, many of whom were contacted, in the first instance, by the Royal Academy whose records of sales at Summer Exhibitions go back to the time of Tunnicliffe's early submissions. We should like to thank the Academy for their generosity in helping us in this way. Other collectors have been contacted through the various galleries noted below and with the help of Tunnicliffe's admirers throughout the country. In view of the value of these paintings we have not given the owners names, but we are deeply indebted to them for their kind co-operation.

We should like to record our thanks to Clive Adams and the Mostyn Gallery, Llandudno; Brian Booth and the Tryon Gallery, London; Richard Hagen Fine Paintings of Broadway, Worcestershire; Peter Keyser and the Keyser Gallery, Malmesbury; John Noote Fine Paintings and the Picton House Gallery of Broadway, Worcestershire, for helping us to locate several of the pictures in this book.

Various pictures are owned by public museums, libraries and art galleries namely:–

The Grundy Art Gallery (Shoveler Duck)
Gwynedd Library Service (Herons Cove)
The National Library of Wales (Buzzard and Dryslwyn Castle)
Newport Museum and Art Gallery (The Rivals and Geese in Hoarfrost)
Nottingham Castle Museum (Gaggle at the Bar)
Shipley Art Gallery (The Ninth Wave)
The Ulster Museum (Lanner Falcon and Whoopers Alighting)

and are reproduced here with their permission.

As noted in the text, Charles Tunnicliffe's sketchbooks were bought by the Anglesey Borough Council and we are grateful to Glyn Jones (and to Dr. Peter

Cannon-Brooks of the National Museum of Wales who granted us access whilst they were in his safe keeping) for permitting us to photograph the selection reproduced herein.

The composition sketches are owned by Bunny Bird, of London, we thank him for bringing them to our attention and for allowing us to photograph them for the purposes of this book.

Others whose help has been invaluable in various ways include David Burnett, Robert Gillmor, Nicholas Hammond and the R.S.P.B., Janice Robertson, Mike Fear, Mary Axon, Graham Sadd, Tony Bently, Don Williams, Ken Williams and the Penrhos Nature Reserve, Messrs. Sotheby & Co., Messrs. Christie Manson and Woods Ltd., Julian Royal and Royles Publications Ltd., Charles Howell and the Medici Society Ltd.

Extracts of text from *Bird Portraiture* are reproduced by permission of Macmillan Publishing Co., New York, and from *A Sketchbook of Birds* by permission of Victor Gollancz Ltd. The appreciations of *Mereside Chronicle* and *Shorelands Summer Diary* given in the 'further reading' section of this book are reproduced from *Tunnicliffe's Countryside* with the permission of Ian Niall.

The publisher is particularly thankful to the members of his own family who have always taken a lively interest and finally to Noel Cusa and his wife Mary; for their hospitality, for taking over the considerable task of sifting the raw material and, in collaboration with Nigel Partridge, for producing this tribute to a great artist.